EINSTEIN'S VIOLIN

The Love Affair between Science, Music, and History's Most Creative Thinkers

Douglas Wadle

Copyright © 2022 Douglas Wadle.

All rights reserved. No part of this book may be used or reproduced by any means, graphic, electronic, or mechanical, including photocopying, recording, taping or by any information storage retrieval system without the written permission of the author except in the case of brief quotations embodied in critical articles and reviews.

This book is a work of non-fiction. Unless otherwise noted, the author and the publisher make no explicit guarantees as to the accuracy of the information contained in this book and in some cases, names of people and places have been altered to protect their privacy.

Archway Publishing books may be ordered through booksellers or by contacting:

Archway Publishing
1663 Liberty Drive
Bloomington, IN 47403
www.archwaypublishing.com
844-669-3957

Because of the dynamic nature of the Internet, any web addresses or links contained in this book may have changed since publication and may no longer be valid. The views expressed in this work are solely those of the author and do not necessarily reflect the views of the publisher, and the publisher hereby disclaims any responsibility for them.

Any people depicted in stock imagery provided by Getty Images are models, and such images are being used for illustrative purposes only. Certain stock imagery © Getty Images.

ISBN: 978-1-6657-1779-3 (sc)
ISBN: 978-1-6657-1781-6 (hc)
ISBN: 978-1-6657-1780-9 (e)

Library of Congress Control Number: 2022900955

Print information available on the last page.

Archway Publishing rev. date: 02/18/2022

Für Elise

and

Samuel Collins

To Music

Music: Breathing of statues. Perhaps:
silence of paintings. You language where all language
ends. You time
standing vertically on the motion of mortal hearts.

—*Rainer Maria Rilke*

CONTENTS

Acknowledgments xi

Overture: Accidental Sagacity 1
C: Introductions 18
G: Laureates 26
D: Quadrivium 36
A: Descent with Modification 50
E: Epiphany 57
B: Disorder and Distraction 68
F♯/G♭: Rausch 79
D♭: Atheoretical Cartography 87
A♭: Biomimicry 102
E♭: Semiotic Metaphor 109
B♭: Resonance 128
F: Synesthesia 136
C: Apophany 145
Finale: Sagacity 155

Bibliography 159
Image Attributions 195
Notes 199

ACKNOWLEDGMENTS

Writing a book is a phenomenal way to realize how interconnected the world is and how dependent we are upon our family, friends, and colleagues. I am fortunate to be particularly blessed in this regard.

First and foremost, I want to give a huge thank-you to my wife, Kristin. I am incredibly lucky to have had such a loving and supportive partner for the past twenty-three years. As I have spent more time writing—early mornings, evenings, weekends—she has never complained or wondered (out loud, at least) whether it was an appropriate use of my time. She deserves extra credit for tolerating the numerous stacks of books, papers, and notes that are inconveniently strewn throughout the house.

My children are a tremendous source of emotional and motivational inspiration. This includes my daughter's music, of course, but also her kind and patient soul. My son's loving inquiries as to the status of my writing have helped pushed me to the finish line. "How's the book coming?" "Are you done yet?" I hope I never disappoint that curiosity!

My parents, Michael and Julia, selflessly dedicated themselves to their children's welfare. My father believed that buying books is never a waste of money, and my mother got me started in music from a young age. It undoubtedly sowed the seeds of later interest, though, regrettably, I didn't take advantage of it at the time. Nonetheless, I am grateful for its unseen benefits. As this book details, the music that was sung and played in my childhood home undoubtedly fostered a curiosity and love of learning that continues to make life endlessly fascinating.

My sister, Ali-Oop, graciously served as a capable and critical sounding board. I highly respect her opinions and judgment. Her husband, the artist EMC, though he would have preferred more rock and roll in this piece, was a source of constructive and counterbalancing right-brained ideas.

My good friend Erik Petersen listened, and made thoughtful suggestions in response to, many hours of problems and ideas on our trail runs through the mountains of Montana. Megan Regnerus skillfully and thoughtfully edited early drafts of this book and encouraged me to pursue it further. Scott McMillion helped get me started down this road when he published my first magazine article in the *Montana Quarterly*. Loma Karklins at the Cal Tech Library was gracious and helpful, and Lisa Black at Princeton University Press helped navigate the farrago of permissions required to quote Albert Einstein.

I am particularly indebted to Carrie Krause and Gregory Young, two talented musicians and educators who read early drafts and helped fact-check my music theory. They were generous with their time and stories and gave me many suggestions that vastly improved the final product. Carrie is a committed champion for baroque music in a rural state. She is also my daughter's violin teacher.

Finally, it is important to acknowledge all the wonderfully creative writers whose works I have cited throughout this book. We are fortunate to have such a treasure trove of insightful and intelligent books on music, science, and mathematics. Their labors of love help make this world a better and more interesting place.

OVERTURE:
Accidental Sagacity

I AM NOT A PROFESSIONAL MUSICIAN, MUSICOLOGIST, COMPOSER, OR MUSIC theorist. I dabble on the violin, but my musicianship is far from noteworthy. It might even resemble that of Sherlock Holmes, whose melancholy wailings and exasperating solos drove Watson nuts—or, better yet, the sound made as he scraped carelessly at the fiddle thrown across his knee. I was a middle school piano lesson dropout who, like many, now regrets that lack of commitment. The violin has long been my favorite instrument. Maybe it's partly genetic, like the melodies inherited by some bird species. My grandfather, great-grandfather, and great-great-grandfather all played the violin—two of them quite seriously. I didn't learn to play until after medical school. When Grandpa Paul died, we rescued his old violin from the attic where it had languished since his hands had become too arthritic to play. One of the f-holes was partially mouse-eaten, and the sound post shook loosely inside. Still, after a little renovation, it retained a nice, warm sound. A German Guadagnini copy, it was made around the time Arthur Conan Doyle first wrote about Sherlock Holmes.

In contrast to my own adolescent apathy, my daughter has

steadfastly put in her hours. Since picking up her first 1/64-size violin at age four, she has developed an aptitude for both music and mathematics—despite her parents' lack of talent here. Over the ensuing years, I witnessed her intellectual development with curiosity. I couldn't help but wonder whether her musical training had encouraged the development of these skills.

Confession made, let me now explain that this is a book about much more than *just* music. In one sense, it could be seen as music appreciation for the scientist—or science appreciation for the musician. It is about creativity and serendipity, and about the musical overtones present in scientific insight. It is about the use of music as a tool for activating the magically creative parts of our brain. And by *our* brain, I mean every one of us, not just the prodigies and geniuses, because music stretches its tendrils into the very evolutionary and neurobiological foundations of human thought.

Two characteristics drive most of music's cognitive benefits. First, it coordinates the right and left sides of the brain to work together. It is a common misconception that music and other creative endeavors are right-brained activities. Both sides do have specialized functions, but neither side works in isolation. It's like working on a team with great collaboration as opposed to an antagonistic one. Brain power is exponentially more powerful when the two hemispheres work in tandem. Second, music creates a vast mesh of connectivity within the brain. The brain is like a microcosm of the World Wide Web, encompassing many discrete networks connecting functionally related, but distant, regions of the brain to one another.

Researchers have identified and named a variety of these networks, including the visual, default mode, and salience networks, to name a few. Music not only helps expand and upgrade the wiring in these preexisting networks but creates entirely novel networks based upon our auditory biography.[1] This is a rapidly changing field, and technological innovations like functional MRI, PET scans, and EEG (electroencephalography) have dramatically increased our knowledge about music's impacts on the brain. I will shy away from too much technical detail, but I hope by the end of this book you will have a better appreciation for the depth and importance of these ramifications.

Abstract pattern recognition is as much a part of scientific thought as it is for mathematical and musical thought. Epiphany is the process of seeing this latticework where it had previously been overlooked. It stems from the Greek *epi*, meaning upon or before, and *phainein*, meaning to show. In other words, it means to reveal something before others—to make salient. As a corollary, we should also consider its opposite. Our brains are so attentive to pattern recognition that they sometimes lead us astray. Apophenia is seeing patterns where none truly exist. By adding the prefix *apo* instead, implying "away, off, or apart from," it now distances us from the truth. These convincing delusions have driven innumerable conspiracies and fallacies throughout human history.

Science uniquely trains the art of thinking. A PhD is, after all, a Doctor of Philosophy. Science shows us not just how to understand the world but how little of it we actually do understand. Scientists are bombarded by a firehose of information and misinformation. It would be impossible to do this work without the benefit of systems to reconcile these discrepancies. A mental toolbox is necessary to avoid the traps and biases that can lead us down a primrose path. Pattern recognition and problem-solving, intriguingly, are also the means by which music interacts with our minds. It is not a coincidence that music has served as the modus operandi of many successful thinkers, including countless Nobel laureates whose stories will be told in the coming pages. Albert Einstein is the chief protagonist in this story. His eccentricities and rebellious nature are well known, but few appreciate how much music played a role in his thought process. Einstein was not alone in embracing music as a tool for insight. It is almost universal enough to be considered a rule. Many of Einstein's contemporaries also successfully integrated art and science. Max Planck, Werner Heisenberg, Paul Ehrenfest, Louis de Broglie, Marie Curie, and the trio of Otto Frisch, Otto Hahn and Lise Meitner exemplify this point.

FIGURE 1: "THE MOST INTELLIGENT PHOTOGRAPH EVER TAKEN" AT THE 1927 SOLVAY CONFERENCE. ALBERT EINSTEIN (FIRST ROW, FIFTH FROM LEFT), MAX PLANCK (FIRST ROW, SECOND FROM LEFT), PAUL EHRENFEST (TOP ROW, THIRD FROM LEFT), LOUIS DE BROGLIE (SECOND ROW, THIRD FROM THE RIGHT), MARIE CURIE (FIRST ROW, THIRD FROM LEFT), WERNER HEISENBERG (TOP ROW, THIRD FROM RIGHT) AND ERWIN SCHRÖDINGER (TOP ROW, SIXTH FROM LEFT).

The correlation between music and creative genius goes back to Leonardo da Vinci and Galileo and, further yet, to Plato and Pythagoras. Like a *da capo* that takes us all the way back to our prelapsarian beginnings, we see the impact of at least forty thousand years of human musical history. From this perspective, the twenty-five hundred years since Pythagoras is almost a blink of the eye. Music has been with us all this time. It is ubiquitous: there is no society on Earth without some form of music. And it is essential: it stirred emotional attachment and lubricated group dynamics in our prehistorical ancestors, enabled communication well before spoken language existed, and may have even helped our ancestors successfully hunt.

Composers have their own unique brand of creative genius. Igor Stravinsky believed his music was divinely inspired. Johann Sebastian Bach, invoking God's inspiration, famously inscribed his scores with J. J. (*Jesu juva*—help me Jesus) or S. D. G. (*soli Deo gloria*—to God alone the glory). Some composers, like Claude Debussy, literally dreamed their music, while others, like Dmitri Shostakovich and Pyotr Ilyich Tchaikovsky, hallucinated theirs. Highlighting the fine distinction between genius and insanity, a disproportionate number of composers, like Robert Schumann and Friedrich Nietzsche, eventually lost contact with reality altogether. The synesthetes Joseph Maurice Ravel and Alexander Scriabin *drew* their music as sound paintings in the colors they visualized when certain notes or keys were played. Bach built a web of intricate patterns into his music. He wrote, played, and improvised in unfathomably complex motifs. It is primarily this structural framework of music that furnishes its metaphorical relationship with science and mathematics.

The philosopher Karl Popper, himself a pianist, emphasized that scientific breakthroughs don't usually extend from simple observation. Instead, they arise from strokes of insight and serendipity. Alexander Fleming's discovery of penicillin in 1928 is often held up as the epitome of epiphany. In fact, it took more than six years before his lucky observation really bore fruit. In 1922, suffering from a head cold, Fleming's sneeze contaminated a petri dish of bacteria. Tossing it aside, he presumed the culture ruined. A few days later, though, cleaning up his desk, he noticed that many of the bacterial colonies had died. Curiosity piqued, he investigated further. His work eventually led to the isolation of the protein lysozyme, part of our mucus and tears and part of our innate armamentarium against would-be microbial invaders. He spent the following years looking for other molecules with similar properties. In 1928, after leaving a petri dish of bacteria by an open window, he stumbled upon the compound produced by a penicillium mold. Primed by his experience with the sneeze, he once again had the foresight to capitalize on this second lucky contamination. The result of his hard work, planted by serendipity and fertilized by curiosity, earned him the 1945 Nobel Prize. With typical modesty, he confessed, "Nature makes penicillin, I just found it; one sometimes finds what one is not looking for."[2] Genius loves serendipity.

The key component of serendipity is the *noticing*. At a Pennsylvania commencement address, Dr. Fleming spoke about its importance. "Pasteur's often quoted dictum that fortune favors the prepared mind is undoubtedly true," he reiterated, "for the unprepared mind cannot see the outstretched hand of opportunity."[3] The synergism of observation and curiosity, or lack thereof, is no better illustrated than by one of Fleming's contemporaries. Just over a decade before Fleming's sneeze, Sir John M'Fadyean at the Royal Veterinary College, under almost identical circumstances, threw out the baby with the bathwater. He exhorted his students:

> Look at this as a warning against delay. Before I could cover the petri dish a spore of some kind of fungus floating about in the air alighted about the culture medium. It will be noticed that the growth from this spore of a parasitic fungus apparently akin to the common mold, has also flourished, and that the growth of this fungus has had a marked deleterious effect on the growth of the staphylococci, either owing to exhalation or addition of something to the culture medium. In fact, the growth of the colonies immediately

adjacent to the fungoid growth appears to have been completely inhibited. The only thing to do now is to throw the culture away.[4]

Dr. M'Fadyean *saw* but lacked the curiosity to *notice*. As Abraham Flexner wrote in *The Usefulness of Useless Knowledge*, "Curiosity, which may or may not eventuate in something useful, is probably the outstanding characteristic of modern thinking. It is not new. It goes back to Galileo, Bacon, and to Sir Isaac Newton, and it must be absolutely unhampered."[5] Michel Foucault added the importance of escaping the tedium of routine. Curiosity, he said, evokes "a readiness to find strange and singular what surrounds us; a certain relentlessness to break up our familiarities."[6]

As Flexner pointed out, serendipity has been with us a long time. But we didn't have a name for it until about 250 years ago. English writer Horace Walpole loved to neologize. He coined such descriptive words as *beefy, somber, nuance*, and *souvenir*.[7] After a coincidental experience, he wrote to his friend Horace Mann in 1754: "This discovery, indeed, is almost of that kind which I call *Serendipity*, a very expressive word."[8] Describing a "silly fairy tale," he explained:

> As their Highnesses travelled, they were always making discoveries, by accidents and sagacity, of things which they were not in quest of: for instance, one of them discovered that a mule blind of the right eye had travelled the same road lately, because the grass was eaten only on the left side, where it was worse than on the right—now do you understand Serendipity?[9]

Accidental sagacity precisely captures the essence of creative insight. It subtly distinguishes between luck and coincidence and adds the wisdom to notice and the curiosity to pursue.[10]

The "silly fairy tale" was *The Three Princes of Serendip*, an ancient Persian fable first brought to the attention of the Western world in 1719 by the chevalier de Mailly.[11] Voltaire borrowed heavily from *The Three Princes* when he wrote the novelette *Zadig, or the Book of Fate* in 1747. Zadig was a young and idealistic philosopher who became disenchanted by the superficialities of life in Babylon. He moved to a remote area on the Euphrates River, where he studied the local flora and fauna, honing his powers of observation until he could see "a thousand differences where others could see only uniformity."[12]

Thomas Henry Huxley, known as Darwin's bulldog for his advocacy of Charles Darwin's theory of natural selection, preached the power of observation in the advancement of scientific theories. Describing what eventually became known as the method of Zadig, he wrote, "What, in fact, lay at the foundation of all Zadig's arguments, but the coarse, commonplace assumption ... that we may conclude from an effect to the pre-existence of a cause competent to produce that effect."[13] He dubbed this "retrospective prophecy": the ability to *predict* past events from current observations.

The legendary Johns Hopkins physician Sir William Osler required *Zadig* be read by all his medical students.[14] He taught that "the whole art of medicine is in observation."[15] His students soon realized, though, that "there is no more difficult art to acquire."[16] Osler encouraged his students to employ "the method of Zadig, that of making deductions or inferences from inconspicuous features."[17]

This was the golden age of medical examination, full of great names like Oliver Wendell Holmes, William Stewart Halsted, William Osler, and, across the pond, Joseph Bell. Bell was a surgeon at the University of Edinburgh, served as Queen

Victoria's personal surgeon during her travels to Scotland, and was a legendary diagnostician.[18] His mentor, the celebrated surgeon James Syme, taught his students to "learn the features of a disease or injury as precisely as you know the features, the gait, the tricks of manner of your most intimate friend."[19] Bell inherited this legacy of astute observation and curiosity. Like Osler, he traced this back to Voltaire:

> Voltaire taught us the method of Zadig, and every good teacher of medicine or surgery exemplifies every day in his teaching and practice the method and its results. The precise and intelligent recognition of minor differences is the real essential factor in all successful medical diagnosis.[20]

As Zadig learned the minutest details of his surroundings, so too must the physician "cultivate the habit of noticing the little apparent trifles."[21]

One of Bell's medical students was profoundly influenced by these lessons. Arthur Conan Doyle graduated from the University of Edinburgh's school of medicine in 1881. He envied that Bell "often learned more of the patient by a few quick glances than I had done by my questions."[22] After graduation, Doyle served a year as ship doctor for the Royal Navy, during which time he visited both the west coast of Africa and the Arctic. Upon his return, he opened a practice in Southsea, where he proceeded to forge Sherlock Holmes from the essence of Dr. Joseph Bell. Bell reveled in the attention he garnered as Holmes's model. He wrote that Doyle's medical training had taught him how to properly observe, and he praised Holmes as "half doctor, half virtuoso."[23] Doyle wrote sixty stories between 1887 and 1929. They brim with medical references, including descriptions of sixty-eight diseases, forty-two of his real patients, thirty-eight different doctors, twenty-two drugs, six hospitals, three medical journals, and two medical schools.[24] And, of course, Holmes's intellectual foil, Dr. John Watson.[25]

Just as Doyle emerged from the tradition of Osler and Bell, the modern detective novel descended from *The Three Princes* and *Zadig*. They first influenced French writer Émile Gaboriau when he created Monsieur Lecoq. They did the same for Edgar Allen Poe's conception of C. Auguste Dupin. Lecoq and Dupin both employed the method of Zadig, combining creativity and scientific knowledge to reveal remarkable feats of observation and deduction. In the *Mystery of Marie Rogêt*, Dupin, echoing the Bell and Zadig quotes above, maintained that the "vast, perhaps the larger, portion of truth arises from the seemingly irrelevant."[26] Another time, in the *Murders in the Rue Morgue*, he reasoned, "Truth is not always in a well. In fact, as regards the more important knowledge, I do believe that she is invariably superficial. The depth lies in the valleys where we seek her, and not upon the mountain-tops."[27] It was from the fertile soil in these valleys that Sherlock Holmes germinated.

The most enduring literary characters often possess a sense of historical presence, and Holmes is no exception. He is sometimes misunderstood as a model of studied perfection. Flawless characters, however, are boring and easily forgotten. Holmes had no shortage of foibles. He was addicted to cocaine, for one thing. This was a common malady in the halcyon days of early pharmacology, before we felt the other sharp edge of its sword. Not coincidentally, Doyle's contemporary Dr. William Stewart Halsted, who popularized cocaine's use as an anesthetic, also became addicted during the course of his research. Holmes was frequently moody and morose. He smoked like a chimney and referred to a particularly challenging problem as a "three pipe problem."[28] Other times, he was arrogant and self-righteous. He disparaged his predecessors, calling Dupin, "a very inferior fellow," and Lecoq, "a miserable bungler."[29] These idiosyncrasies, though, are the very traits that fascinate us. "We may admire Lecoq," wrote Bell, "but we do not see ourselves in his shoes."[30]

Holmes remains to this day the embodiment of observation and logic. The method of Holmes was essentially the method of Zadig, "founded upon the observation of trifles."[31] He emphasized that "it is of the highest importance in the art of detection to be able to recognize, out of a number of facts, which are incidental and which vital."[32] As infants, we become adept at filtering out unnecessary details in order to focus only on what's most important.[33] A shocking amount of salience flies under our radar. Unmasking this filter as an adult takes assiduous attention. Holmes told Watson, "The world is full of obvious things which nobody by any chance ever observes."[34] Another time, he lectured him, "I see no more than you, but I have trained myself to notice."[35] A trained observer notices things that are different enough to arouse their curiosity. It is curiosity and imagination overlaying a bedrock of scientific training that places Holmes in rarefied company. It is genius.

Arthur Schopenhauer defined genius as the ability to "see what everybody else has seen, and think what nobody else has thought."[36] This is exactly what Holmes does in *Silver Blaze*. "See the value of imagination," said Holmes, explaining how he solved a case that Inspector Gregory could not, "it is the one quality which Gregory lacks. We imagined what might have happened, acted upon the supposition, and find ourselves justified."[37] Einstein would have agreed with Holmes. "Imagination is more important than knowledge," he said. "Knowledge is limited. Imagination encircles the world."[38]

Curiosity could be seen as the driving engine of Einstein's genius. As he confessed to his first biographer, Carl Seelig, "I have no special talents, I am only passionately curious."[39] What he called his "murderous curiosity," though, was something to behold.[40] Curiosity alone, however, is not sufficient; for the greatest impact, it must coexist with creativity. Without creativity, curiosity is quotidian; and without curiosity, creativity is mere amusement. The musicians and thinkers in this book are not just creative in an *artsy* kind of way. They are creative in a *where-did-that-come-from?* kind of way. Difficult to fathom, out-of-the-box thinking shows just how much of the observable world goes unperceived. "There are more things in heaven and Earth," wrote the Italian physicist Carlo Rovelli, "than have been dreamed of in our philosophy."[41] The window into this dreamworld is propped open by genius. These are not merely superintelligent people. More importantly, they see patterns and make connections that are beyond comprehension by the rest of us.

FIGURE 2: ALBERT EINSTEIN IN 1921, THE YEAR HE FIRST VISITED THE UNITES STATES AND A YEAR BEFORE HE WON THE NOBEL PRIZE.

As a physician, I strive to solve problems as effectively as Sherlock Holmes. Unlike the crimes in Doyle's books, however, medicine contains few absolutes. There are no unique sets of clues, so medical problems are rarely soluble by logic alone.[42] Further evidence—or testing—is nearly always necessary. Huxley's "retrospective prophecy" must be employed to put all the clues in proper perspective—that is, to ascertain what past events may have led to our current observations. Real patients present a complex assortment of motives, symptoms, and lifestyles. The art of medical diagnosis requires creativity and ingenuity. It is like putting together a jigsaw puzzle, but this puzzle is missing pieces and has a hodgepodge of pieces from other puzzles mixed in.[43] In the end, it is pure pattern recognition.

Perhaps this is one reason why so many physicians also play music. French composer Hector Berlioz studied medicine until he could no longer stand the revulsion of the dissection laboratory.[44] Aleksandr Borodin was a physician and served a year as a military hospital surgeon before pursuing his career as a chemist and part-time composer.[45] Cardiologist Richard Bing was another part-time composer. Virginia Apgar, whose scoring system attends the birth of every child in the United States, was a violinist and even made a complete set of instruments for a string quartet.[46] The famous French physician René Laënnec was trained as a flautist. It is with his story that we will return to serendipity.

Laënnec is best remembered for his invention of the stethoscope. At the time, in the early 1800s, physicians listened to the heart and lungs by placing an ear directly against the chest of the patient. This presented difficulties with obese patients. Or, for the prim Laënnec, the case of a particularly buxom young lady "labouring under general symptoms of diseased heart."[47] While puzzling over this woman's case, he went on a walk. Passing by some boys playing with a long piece of hollow wood, he noticed that while one scratched with a pin at one end, his friend heard the sound transmitted at the other.[48] Feeling the warm flush of epiphany, Laënnec returned to his patient. He rolled up a sheet of paper into a cylinder and could now listen—at an appropriate remove—to the young woman's chest. "I was surprised and elated," he wrote, "to hear the beating of her heart with far greater clearness than I ever had with direct application of my ear."[49]

Laënnec was motivated to study diseases of the chest after his mother died of tuberculosis when he was only five years old. His brother and uncle also died of the disease. Laënnec would eventually suffer the same fate. After his mother died, Laënnec soothed himself with music. He became a talented flautist and developed a skill carving flutes that later aided his design of a wooden stethoscope tube. His musician's ear enabled wonderfully vivid descriptions of the lung sounds that could now be heard with his new invention. A new vocabulary was needed.

His poetic descriptions of phenomena like "pectoriloquy" and "egophony" evoked a soundscape still used by physicians to this day. A tubercular cavitation in the lungs, he wrote, "consists of a peculiar sound which bears a striking resemblance to that emitted by a cup of metal, glass, or porcelain, when gently struck by a pin, or into which a grain of sand is dropped." Other conditions created a tinkling, like "a gnat buzzing within a porcelain vase." Laënnec's musical background equipped him to notice the most subtle sound characteristics. Some were "sonorous," while others were more "sibilant." The sonorous variety reminded him of a cooing wood pigeon: "This resemblance is sometimes so striking, that we might be tempted to believe the bird concealed under the patient's bed." The sibilant variety, on the other hand, could be either "flat or sharp." It resembled, to his ear, "the chirping of birds," or even "the note of a violoncello." Bronchitis irritated his ears like "the prolonged scrape of the bow on a large violoncello-string." A heart murmur was "like the sighing of wind through a keyhole," while a palpated pulse revealed the subtle "vibration ... of a fiddle-string when touched."[50] He once even transcribed the "melody" of the arteries onto a musical staff.[51] Years later, composer John Cage described this same melody humming in his ears while immersed in the anechoic chamber at Harvard University.

Laënnec's serendipitous insight at the wooden log and his creativity to extrapolate it into something completely new inspired one of the most important inventions in medical history. The stethoscope is now a ubiquitous symbol of medicine. One cannot imagine a physician without one around her neck. Laënnec's musical background had a significant influence on this insight.

Music happens to be a frequent driver of insight—a serendipity stimulant. By forcing the creative right brain to communicate with the analytical left side, music greases the groove between the conscious and the unconscious, between the logical and the haphazard, and between the cautious and the reckless. The inordinate number of musicians who have made major impacts upon their respective fields seems far greater than mere coincidence would suggest. It highlights the syllogism of science and music, of pattern recognition and creativity. Classical music has particularly complex structures and abstract patterns. This drives its cognitive power. American pianist Charles Rosen scorned music that exploited "voluptuous noise" at the expense of a foundational bony structure.[52] For him, classical music's drama—its power—*is* its structure.[53] Pattern and structure recruit the problem-solving, pattern-recognition software of the right side of the brain. Better yet, they get the two halves of the brain working together.

These concepts are exactly why music can help every one of us think more creatively and insightfully. Researchers have now demonstrated the powerful impact that music has on brain plasticity and cognitive reserve. They have found that the simple act of learning an instrument can decrease the risk for dementia, improve prognosis after a stroke, and mitigate the cognitive patterns of dyslexia. While the greatest benefits of music occur in children, it is now clear that they also translate to the adult brain. Learning an instrument at any age has proven advantageous, *regardless of whether we are any good at it.*

What Einstein called "the happiest thought of my life,"[i] his theory of relativity, came to him in a fit of music-inspired epiphany.[54] He told Shinichi Suzuki, the founder of the famous Suzuki Method of music instruction, that music inspired his mathematical intuition.[55] When his theory of general relativity was first published, it was so poorly understood that a joke went around concluding that only three scientists in the world understood it.[56] Nonetheless, his colleague (and sometimes duet companion) Max Planck estimated, "If Einstein's theory [of relativity] should prove to be correct, as I expect it will, he will be considered the Copernicus of the twentieth century."[57] Indeed, it has become vital to every one of our lives: Global Positioning Systems (GPS) would quickly become obsolete without it. Orbiting satellites move so fast that it causes a shift in their clocks by about 38 milliseconds per day. That may not sound like much, but without Einstein's relativity equations, GPS would lose seven miles of accuracy every day.[58] Instead, it can pinpoint your location to within a foot.[59] How's that for serendipity? Einstein discovered patterns and relationships with almost no foreseeable application that now contribute to systems he couldn't have even imagined.

Pythagoras, two and a half millennia ago, created a music theory grounded in the ratios of stringed instruments: 2:1 for the octave, and 3:2 for the fifth. Russian composer Nikolai Diletsky, in his 1679 *Grammatika*, introduced the circle of fifths (see figure 3) as a logical outgrowth of this Pythagorean theorem.[60] As we step up by a fifth from the C note, four keys on the piano, we arrive at G. Another fifth gets us to A, then to E, B, G♭, D♭, A♭, E♭, B♭, F, and finally back to C again. The note

---

[i] Einstein used the German, *"glücklichste Gedanke meines Lebens."* Walter Isaacson noted that although it is usually translated as "happiest," it could also mean "most fortunate." (Isaacson, *Einstein*, 145.)

that began our scale—C, in this case—is called the tonic, and it is the note from which all the others are referenced. The circle of fifths is vital for musicians to understand. It can be used, for example, to develop chord progressions or to modulate to another key. It also helps understand how many sharps or flats are in a given key. The violin's strings, for example, are tuned a fifth apart—G, D, A, E—one sharp, two sharps, three sharps, and four. I promised not to get bogged down in theory. I mention this now only because each chapter in this book represents a component of this circle. Like the circle of life, everything leads back from whence it came, returning again and again to the tonic. And like everything else in this strange world of ours, nothing—including the theory behind the circle of fifths—is as simple as it seems.

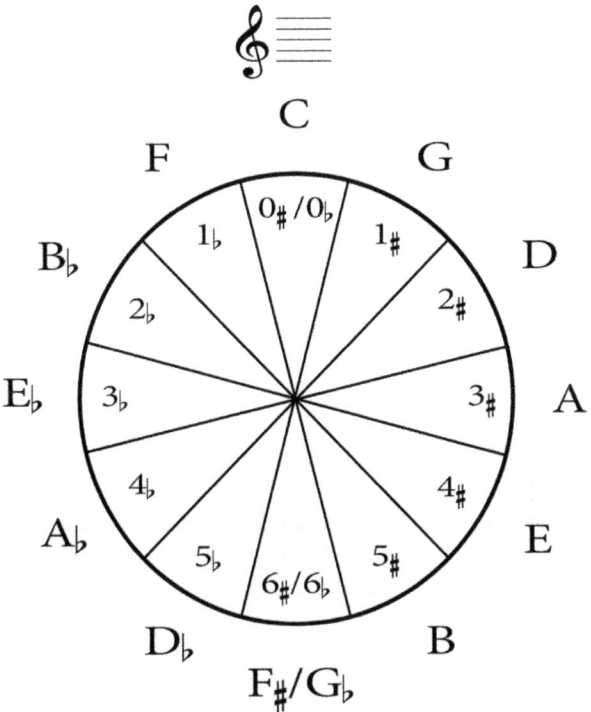

FIGURE 3: THE CIRCLE OF FIFTHS.

C:

Introductions

One evening while Albert Einstein sat at home talking with his landlady, the muted tones of a piano sonata by Wolfgang Amadeus Mozart resonated through the walls. Upon inquiry, his landlady reported that an old woman lived next door and taught piano. Impulsively—and to his landlady's horror—he grabbed his violin and rushed over to the neighbor's home without so much as putting on a collar or tie. The neighbor lady, startled as he burst into her room, gaped in astonishment from the piano. Einstein pled with her to keep playing, and soon, the two were weaving melody and harmony together in a lovely duet.[1]

Einstein had strong opinions about the classical composers. Bach and Mozart were clearly his favorites. He maintained that their music had similar architectural structures to his favorite scientific theories. It was as if they were plucked straight from the universe rather than merely composed.[2] He also liked Franz Schubert for his unequaled expression of emotion. Richard Strauss, on the hand, had no inner truth. Wilhelm Richard Wagner lacked structure, Felix Mendelssohn was banal, and George Frideric Handel was too shallow.[3] Although Einstein frequently played Beethoven, he considered his music "too personal, almost naked. Give me Bach, rather, and then more Bach."[4] After he witnessed a particularly inspiring performance in November 1924, he wrote a letter to the German conductor and composer Siegfried Ochs:

> You brought out Beethoven's profundities as virtually no one else could. I heartily thank you for this experience that you conveyed to me ... God is not standing there but Beethoven, the human being. But he can stand it, since that fellow is a giant. And yet, the lucidity and self-expression that Bach and Mozart achieved do seem to me to come closer not just to me but, I believe, to the sensitivity of our time.[5]

Whereas Beethoven's music was merely a marvelous expression of creativity, "Mozart's music is so pure and beautiful that I see it as a reflection of the inner beauty of the universe," he wrote.[6] Einstein felt a close emotional and intellectual kinship with Mozart, whose music mirrored his own beliefs about the universe's secret hidden beauty.

Einstein implicitly believed there was a hidden underlying structure to the universe. Like Mozart's music, it was out there just waiting to be discovered. The same purity he found in Mozart's music he sought in the heavens. Scientists in the early twentieth century were notably not focused on theories of beautiful simplicity. They were kept busy performing mental acrobatics in order to explicate conflicting theories. They still believed in the ether, for one thing. Originally conceived by Aristotle, ether was thought to fill the empty space of the heavens. It was a handy invention to account for the fact that light waves cannot travel through a vacuum. James Clerk Maxwell and Michael Faraday had only recently unified electricity and magnetism, but Newtonian gravitational law still ruled physics. Einstein chafed at the need for different formulae and concepts in each separate field within physics. It was ugly. And it was anathema to what he believed the Lord would have created. In homage to their groundbreaking unification of electromagnetism, Einstein hung portraits of Faraday and Maxwell in his home office.[7] Einstein sought to do the same thing for all of physics: a magnificent simplification within a single mathematical model. His stubborn belief in this grand architecture, speculated his biographer Walter Isaacson, "made Einstein a rebel with a reverence for the harmony of nature."[8]

Music was no effete indulgence. Rather, it was a tool for creative epiphany. Einstein's son, Hans Albert, once described how music influenced his father's problem-solving process: "Whenever he felt that he had come to the end of the road or faced a difficult challenge in his work, he would take refuge in music and that would solve all his difficulties."[9] A friend reported, "He would often play his violin in his kitchen late at night, improvising melodies while he pondered complicated problems. Then suddenly, in the middle of playing, he would excitedly announce, 'I've got it!'"[10]

FIGURE 4. EINSTEIN PLAYING MUSIC WITH THE GERMAN MATHEMATICIAN ADOLF HURWITZ AND HIS DAUGHTER LISBETH HURWITZ, CA. 1913.

"Intelligence loves patterns," wrote mathematician Douglas Hofstadter, "and balks at randomness."[11] This certainly described Einstein. The same could also be said of Sherlock Holmes, who epitomized the usefulness of music in exposing hidden patterns. Music honed his deductive powers, facilitating out-of-the-box thinking. Holmes demonstrated his observational prowess in *A Study in Scarlet*. After their first meeting, he explained to Watson how he instantly knew so much about him:

> Here is a gentleman of a medical type, but with the air of a military man. Clearly an army doctor, then. He has just come from the tropics, for his face is dark, and that is not the natural tint of his skin, for his wrists are fair. He has undergone hardship and sickness, as his haggard face says clearly. His left arm has been injured. He holds it in a stiff and unnatural manner. Where in the tropics could an English army doctor have seen much hardship and got his arm wounded? Clearly in Afghanistan.[12]

Logic simply applies the known facts to smaller questions in order to eventually arrive at the logical answer. Lewis Carroll, in *Through the Looking-Glass*, humorously danced around this process: "'Contrariwise,' continued Tweedledee, 'if it was so, it might be; and if it were so, it would be; but as it isn't, it ain't. That's logic.'" Logic includes both deductive and inductive reasoning. Deductive reasoning goes from the general to the specific. It is the scientific method. It is syllogism: "All men are mortal, Socrates is a man, therefore Socrates is mortal."[13] It also happens to be how mathematicians prove theorems. Inductive reasoning, on the other hand, translates the specific to the general. This is often where scientific theories come from. But it is more dangerous and at risk for overgeneralization. Just because you pulled three pennies out of your pocket does not mean all the coins in your pocket are pennies. Sherlock Holmes described inductive reasoning when he explained, "So all life is a great chain, the nature of which is known whenever we are shown a link of it."[14] It allowed him to see through the farrago of apparently meaningless clues to arrive at an insightful, and often unexpected, conclusion.

Holmes's mental toolbox held many tools. One of these tools was Watson himself, as intellectual foil. Then there was the tobacco pipe, which put him into a meditative state of mind. But the key implement in his toolbox was music—his violin in particular—which helped elucidate patterns hidden behind apparent randomness.

The violin was in some ways simply a stress reliever for the busy Holmes. In *The Red-Headed League*, he sighed and said, "A sandwich and a cup of coffee, and then off to violin-land, where all is sweetness and delicacy and harmony, and there are no red-headed clients to vex us with their conundrums."[15] Music was integral to Holmes's persona. He was particularly enamored by the virtuoso, reflecting his own narcissism within his field. Niccolò Paganini, Pablo de Sarasate, and Wilma Norman-Neruda are depicted throughout his works. In *The Adventure of the Cardboard Box*, Watson and Holmes sat "for an hour over a bottle of claret while [Holmes] told [him] anecdote after anecdote" about the virtuoso violinist Niccolò Paganini.[16] In *The Red-Headed League*, Holmes attended a performance by Pablo de Sarasate at St. John's Hall: "There is a good deal of German music on the program, which is rather more to my taste than Italian or French. It is introspective, and I want to introspect."[17] Watson, also in attendance, described Holmes as enthralled, "wrapped in the most perfect happiness, gently waving his long, thin fingers in time to the music."[18] He surely would have thought highly of more recent prodigies like Jascha Heifetz, Yehudi Menuhin, and Midori Gotō. Einstein and Holmes had this much in common. Einstein gushed over Menuhin after he witnessed the wunderkind's debut in Berlin. He personally played chamber music with the Russian virtuoso Toscha Seidel, as well as the legendary violinist and composer Fritz Kreisler.

FIGURE 5. COMPOSER AND VIRTUOSO VIOLINIST PABLO DE SARASATE.

Over the course of sixty stories, Holmes never once attended a symphonic work. He, like Einstein, preferred solo violin performances. Often these were transcriptions. It was not uncommon in the late 1800s for violinists to play pieces initially written for different instruments. The music scholar Ernest Newman noted that this sometimes served the composer well, and "put the classics in their place."[19] For instance, Antonín Dvořák's *Humoresque* was originally written for piano. It attracted little attention until Fritz Kreisler transcribed it into a violin arrangement, "and the whole world took it to its heart."[20] It was an error on Dvořák's part, Newman claimed, to set it for the piano. He professed similar sentiments about Frédéric

Chopin: "I do not know who began the practice of playing the E-flat Nocturne as a violin solo; but whoever he was he deserves the thanks not only of humanity but of Chopin. Many of Chopin's melodies are obviously so much more violinistic than pianistic that one wonders why Chopin himself did not perceive that fact."[21] In *A Study in Scarlet*, Holmes attended a performance by the Moravian virtuosa Wilma Norman-Neruda. "What's that little thing of Chopin's she plays so magnificently: Tra-la-la-lira-lira-lay," he asked.[22] Holmes appreciated what Newman described as "the flexibility of rhythm, the nuance of expression, and the delicacy of melodic modelling that are the secret of the violin and the violin alone among instruments."[23]

Predictably, Holmes owned a Stradivarius. Boring Watson, "he prattled away about Cremona fiddles, and the differences between a Stradivarius and an Amati."[24] Holmes purchased his on Tottenham Court Road for a mere 55 shillings, or less than 3 pounds.[25] What a bargain! Even in the late 1800s they were selling for 300 pounds or more.[26] Nowadays, they sell for millions: In 2011, the Lady Blunt violin (named for Lord Byron's granddaughter) sold for nearly $16 million.[27] That same year, violinist Joshua Bell acquired the 1713 Gibson ex-Huberman Strad for $4 million.[28] It's a small world: violinist Bronisław Huberman, the Huberman of Bell's Strad, was a good friend of Einstein's. He once visited Einstein at his home in Princeton, and it is likely he allowed Einstein to play the mythical instrument.[29] Unfortunately, it was later stolen from Huberman's dressing room at Carnegie Hall. The violin finally resurfaced, many years later, after years of abuse playing in the bars of New York City. Einstein's own violin eventually attained its own mythos and a price tag to match[ii]—selling for nearly the price of a Strad.[30]

Initially, Watson mistook Holmes's violin playing as mere diversion and recreation. Eventually, though, he became acquainted with the manifold moods of Holmes and his violin:

[ii] Einstein's Oscar Steger violin sold at auction for $516,500 in 2018.

> That he could play pieces, and difficult pieces, I knew well, because at my request he has played me some of Mendelssohn's Lieder, and other favourites. When left to himself, however, he would seldom produce any music or attempt any recognized air. Leaning back in his armchair of an evening, he would close his eyes and scrape carelessly at the fiddle which was thrown across his knee. Sometimes the chords were sonorous and melancholy. Occasionally they were fantastic and cheerful. Clearly, they reflected the thoughts which possessed him, but whether the music aided those thoughts, or whether the playing was simply the result of a whim or fancy, was more than I could determine.[31]

Watson eventually recognized the key role music played in Holmes's problem-solving toolbox. One night, Watson reported, "I left Holmes seated in front of the smouldering fire, and long into the watches of the night I heard the low, melancholy wailings of his violin, and knew that he was still pondering over the strange problem which he had set himself to unravel."[32] His violin was a tool in the literal sense of the word as well: an implement, especially one held in the hand, used to carry out a particular function. It brought his creativity to life by removing him from the stubbornly logical left side of his brain.

With this, we return to the tonic: Einstein—and an astounding number of other great thinkers—relied on music to access the innate creativity of the brain's right hemisphere. Why are so many brilliant and creative thinkers also musicians? Let's ask the Nobel Prize winners.

G:
Laureates

MOST PEOPLE ARE AWARE THAT EINSTEIN PLAYED THE VIOLIN, BUT MANY DON'T appreciate how much it contributed to his intellectual development. He began musical training at a young age. His talent was noticed by a music assessor at his school. Criticizing the stiff bowing in many students, the evaluator praised Einstein who "sparkled by rendering an adagio from a Beethoven sonata with deep understanding."[1] In many respects, Einstein did not fit into society's mold. Music may have been one driver of his brash personality and lack of respect for the status quo. The eccentric professor. These now celebrated traits were not always admired. He drove his professors crazy. One Zurich Polytechnic professor even flunked him—in physics, no less! He had to wonder why Einstein was wasting his time with physics when he could instead study medicine or even law. Einstein shot back that he was even less talented in those subjects.[2] The unseen force, lurking like an iceberg beneath the surface, was his exceptional creativity.

Creativity is poorly forecast by standard scholastic curricula, which require rote obedience to the teacher's agenda. Einstein had his own agenda. The unchallenged genius who struggles in school is a common trope. Sometimes it seems to be more the rule than the exception. Thomas Edison, whose teachers called him "addled," admitted that he was always "at the foot of the class."[3] Steve Jobs finished high school with a GPA of 2.6 and then dropped out of college after just one semester.[4] Genius seems to travel with a certain degree of stubborn recklessness.

A deep understanding of music fostered Einstein's creativity. Countless images portray him with his violin. He lovingly called her Lina, and she accompanied him everywhere.[5] He dutifully played for his mother's friends at social gatherings and luncheons. As the underpinnings of World War II brewed in Europe, he spent afternoons playing Mozart with Belgium's Queen Elisabeth. They played together so frequently that her husband, King Albert I, began referring to himself as "the husband of the second violinist."[6] Years later, when the queen's husband and daughter-in-law died in close succession, Einstein wrote consolingly: "Mozart remains as beautiful and tender as he always was and always will be. There is, after all, something eternal that lies beyond the reach of the hand of fate and of all human delusions."[7]

FIGURE 6: EINSTEIN WITH HIS DUET PARTNER, QUEEN ELISABETH OF BELGIUM.

When Einstein moved to America, he celebrated his new home by hosting a recital featuring Joseph Haydn and Mozart. This time he was the second violinist, behind Russian virtuoso Toscha Seidel. A fundraiser for Jewish refugees from Nazi Europe found him performing Bach's Concerto for Two Violins and a Mozart quartet. Another time, after a scientific lecture, he told the audience that it might be more pleasant if, instead of giving yet another speech, he played the violin; this time it was a Mozart sonata.[8] Wednesday nights were sacrosanct, avidly protected from other engagements for chamber music in his home.[9] At Christmastime, when carolers came to his door, he stepped outside in the snow to accompany them on the violin.[10] It is evident, as all these anecdotes depict, how integral the violin was to Einstein's identity.

FIGURE 7: RUSSIAN VIRTUOSO VIOLINIST TOSCHA SEIDEL PLAYED WITH EINSTEIN WHEN THE PHYSICIST FIRST ARRIVED IN AMERICA.

Alfred Nobel set out the conditions for what would later become the Nobel Prize in his last will and testament. The story of his will is itself full of insights into human nature. The inventor of dynamite, Alfred was well known as an egotistical Scrooge. It was his military explosives, though, that gave him a reputation as a war profiteer who grew rich by developing new ways to mutilate and kill. When Alfred's older brother Ludvig died, a French newspaper apparently mistook the latter for the former. They ran an eviscerating epitaph about the "Merchant of Death."[11] The misfortune—or fortune—of witnessing his own dubious legacy drove him to amend his ways. A combination of existential guilt and journalistic error led to the prize now considered the pinnacle of human achievement.

Jacobus Henricus van 't Hoff won the first Nobel Prize in chemistry in 1901 for recognizing that molecules could only be properly understood by considering their three-dimensional structure. It took a great leap of imagination to change the existing paradigm, and he was mocked for it—until later proven correct.[12] Music and poetry were an important part of his life, and he played the pianoforte regularly.[13] Some years earlier, speaking as he took up a professorship in Amsterdam, he drew attention to the disproportionate number of eminent scientists with artistic inclinations. He laboriously reviewed a litany of scientific biographies to arrive at the conclusion that almost all of them possessed a high degree of artistic imagination.[14] Many, including the renowned Isaac Newton, Johannes Kepler, René Descartes, and Gottfried Wilhelm Leibniz, even showed signs of what he called "pathological" imagination.[iii] He would probably have placed the reckless Einstein in that category as well. The renowned "father of the atomic bomb," J. Robert Oppenheimer, certainly did. Upon first meeting Einstein, he exclaimed to his brother, "Einstein is completely cuckoo."[15]

[iii] "For I have noticed that not infrequently, examples of the strangest imaginings, superstitions, belief in spirits, hallucinations, and even insanity appear in the biographies. Newton was always afraid than an accident would occur to his coach and held on to the door at all times. Kepler ... believed that the earth was a reptile ..." (van 't Hoff, Imagination in Science, 16.)

Many years later, a study of Nobel laureates confirmed van 't Hoff's hypothesis: there was indeed a strong correlation between scientific success and music.[16] The correlation was most significant for those who did not merely play for relaxation but integrated music into their work. Einstein's contemporaries alone offer up a full roster of musicians as scientists. Werner Heisenberg, the renowned author of the uncertainty principle, was a concert-caliber pianist. Max Planck, who formulated quantum theory, was a talented pianist and would-be composer. Louis de Broglie's wavelike contributions to quantum mechanics were inspired by his musical background. Ludwig Boltzmann was a skilled pianist, and Walther Nernst invented the technology to amplify musical instruments.[17] Einstein's close friend, the physicist Max Born, played the piano.[18] Another of Einstein's friends, the double Nobel Prize winner Marie Curie, presided over a musical household—her daughter Ève later became a professional pianist.[iv] Paul Ehrenfest dabbled on the piano, developing a lasting bond with Einstein after the two played a Johannes Brahms duet one evening.[19] After another musical evening, Einstein raved to his wife, Elsa, about Ehrenfest's magnificent renditions of Bach.[20] These great men and women certainly played for relaxation and enjoyment, but, to the point of the Nobel study, it also served a greater purpose. They played music when a problem was simmering in their mind.

Planck was a giant in physics at the turn of the century. He was among the first to support Einstein's early work when nobody else understood its implications or its importance. He made his own major contributions to the fields of quantum mechanics through his black body experiments. As we might expect from a fellow musician, he wrote to Einstein in the summer of 1907 that together they might "unify all the forces of nature."[21] The Planck family frequently hosted chamber music

---

[iv] Four people have won two Nobels, but Curie and Linus Pauling are the only people to win a Nobel Prize in two different fields.

parties for musician friends and colleagues. The most common trio performing was Planck, his son Erwin, and Einstein.[22] Schumann and Schubert were among his favorite composers, and he was close friends with virtuoso violinist Joseph Joaquim. At one time Planck had even considered pursuing a career as a composer, but he eventually made the wise decision that there was little future for him there. Nonetheless, music remained a big part of his personal life and professional work.

Planck's student, Lise Meitner, described the pleasures of one of these house parties at her professor's home: "Last night I was at Plancks'. They played two marvelous trios, Schubert and Beethoven. Einstein played violin and occasionally made amazingly naïve and really quite peculiar comments on political and military prospects."[23] Meitner broke many glass ceilings over the course of her career. She was the first female to receive a doctorate in physics from the University of Vienna. Then, with Planck's help, she became the first female science professor at the University of Berlin.[24] Meitner and a duet of Ottos—Otto Frisch and Otto Hahn—made key discoveries about nuclear fission in uranium. During her work with Hahn, "He would whistle large sections of the Beethoven violin concerto, sometimes purposely changing the rhythm of the last movement just so he could laugh at my protests."[25] Hahn surreptitiously aided her last-minute escape from Germany, just as she, a Viennese Jew, was about to be arrested by the Reich. After fleeing to Sweden, her nephew Otto Frisch and she continued the work on fission, collaborating with Hahn from afar. The trio should have shared the Nobel Prize for this work, but it was awarded only to Hahn. Frisch, a skilled pianist and occasional violinist as well, had a very successful career in his own right.[26] Along with Rudolf Peierls, he presented the Frisch–Peierls memorandum, the first exposition describing how an atomic bomb could be constructed from a small quantity of fissile uranium 235. The two also collaborated on the detonation mechanism for the bomb.

The alliance between music and science has persisted well beyond Einstein's era. Many recent Nobel laureates have continued this tradition. Biochemist Thomas Südhof, for example, plays the violin and bassoon. He credited this musical training for his "dual appreciation for discipline and hard work on the one hand, and for creativity on the other."[27] Importantly, he pointed out, "There is no creativity if one does not master the subject and pay exquisite attention to the details, but there is also no creativity if one cannot transcend the details."[28] Particle physicist Donald Glaser played viola with the Cleveland Philharmonic as a young man. Mathematician and physicist Frank Wilczek, a pianist, recognized the unique connection between patterns in music and those in science. "Bach is a great manipulator of patterns," he wrote.[29] Chemist Frances Arnold, another pianist, compared the intricate and beautiful code of life to Beethoven's music.[30]

Figure 8: Dr. Donald Glaser won the Nobel Prize in Physics for his invention of the bubble chamber, a precursor to today's modern accelerators like the Large Hadron Collider.

These highly successful scientists all saw music as a gateway to pattern recognition and creativity. One thing that is apparent from the above list of musician scientists is the predominance of instruments in which the two hands perform different types of movements. Within this group of instruments, there appears to be a particular preference for the piano and violin. While it smells of generalization—the problem of induction—scientists seem to prefer these instruments. This predilection even predated the invention of these instruments, as ancient scientists tended to play the lute and lyre instead of today's violin and piano. In Greece, Pythagoras played both of these instruments as he pondered the harmony of the spheres. Leonardo da Vinci, the quintessential genius and polymath, played the lyre. Music was so intrinsic to his work that he sometimes hired musicians to play while he painted.

Galileo was trained in music from a young age. His father Vincenzo Galilei, a musician and composer, taught him to play the lute. Galileo loved how the instrument mirrored human emotion.[31] Later in life, while under house arrest, his frequent letters to his daughter Celeste often included short musical compositions. He also employed music for a more utilitarian purpose. Because there were no accurate timepieces in those days, rhythm and tempo served as his stopwatch. One of his most famous experiments, on the descent of objects down an inclined plane, he timed by humming tunes.[32] The very roots of Einstein's general relativity were planted by Galileo. He reasoned that a sailor below decks, under calm seas and steady speed, would be unable to tell whether the ship was moving.[33] Velocity, he noted, must be referenced against something: it was relative.

FIGURE 9: THOMAS EDISON AND HIS PHONOGRAPH.

There are, of course, exceptions to the genius musician protagonist. Thomas Edison, brilliant by many measures, was hopelessly stymied by music. The worst of it, though, was his stubborn ignorance of the fact. That he was stone deaf didn't help. He discovered that if he bit down on the wood of the phonograph cabinet, he could *hear* the music through reverberations in his skull.[34] Coincidentally, Beethoven used a similar technique after his own deafness set in: "By clenching a stick in his teeth, holding it against the keyboard of his piano, he could discern faint sounds."[35] For Edison, however, the technique led to an eccentric taste in music. He was intolerant of dissonance, which resonated uncomfortably in his skull. He disliked Mozart, whom he thought pitched too high. Edison sometimes judged music by examining the cylindrical record grooves under a magnifying glass. He despised the wavy grooves of a violin's sonorous vibrato. Certain piano combinations were no less grating. His company once invited Sergei Rachmaninoff to a recording session. Edison was not at all impressed by the resulting groove pattern. He blasted Rachmaninoff: "Who told you you were a piano player? You're a pounder."[36]

Sigmund Freud was another brilliant thinker who was

distinctly unappreciative, to say the least, of music. He was fanatical about controlling emotion, so his aversion to music may have been philosophical as much as it was aesthetic: music was just too arousing. In 1924, Freud wrote *The Economic Problem in Masochism* exploring the Nirvana Principle, a term coined by Barbara Low in 1920.[37] It espoused "reducing to nothing, or at least of keeping as low as possible, the sums of excitation which flow upon [the mind]."[38] As his biographer, Peter Gay, wrote, "Freud's life ... was a struggle for self-discipline."[39] Anything that fought this impulse was kept at arm's length. This instinct seemed to be innate. His sister Anna loved the piano and practiced religiously until Freud demanded that his mother sell the offending instrument. Exploiting his favored-eldest-child role, he effectively abolished music from the home.[40] Maybe it was more than just a suppression of latent desires. That it began in childhood suggests he could have had amusia or some sort of melophobia. His nephew Harry later wrote that his uncle "despised music and considered it solely as an intrusion!"[41] Freud regarded music as *otiose*, fulfilling no purpose. This would have been particularly unusual in Vienna, the city of music, where it played such a prominent social role. Apparently, Freud was not immune to being blinded by his own neuroses. His prejudices regarding music, as we will see, missed the broad side of the barn.

D:

Quadrivium

Plato, following ancient Greek traditions from Pythagoras, Parmenides, and Anaxagoras, understood the importance of music in education. He spelled out music's proper role in education in book seven of his *Republic*. Pupils were first given a strong foundation with the trivium—grammar, logic, and rhetoric. This is where they learned how to think and how to communicate those thoughts in the ancient and original version of the three r's. Once this base was established, their education culminated with the more advanced quadrivium.[v] These four *scientific arts* included arithmetic (pure number theory), geometry (numbers in space), music (numbers in time), and astronomy (numbers in time and space). Pay attention to the definition of music as numbers in time. We will return to this idea. Together, the trivium plus the quadrivium formed the basis of a liberal arts education.

More recently, music has been relegated to luxury status in many Western schools. Like Freud, many short-sighted bureaucrats see music as otiose. The quadrivium has faded into

---

[v] Trivium is the place where three roads meet. Latin *tri* + *via*, or three roads. Quadrivium, likewise, is the place where four roads meet.

historical curiosity. This turn of events would abhor the ancient Greeks, who insisted their pupils sing and play the lyre. Socrates believed that music was the most potent instrument, "because more than anything else rhythm and harmony find their way to the inmost soul and take strongest hold upon it."[1] Plato portrayed music as a counterbalance to the more aggressive nature of the gymnasium. Unopposed, athletics would lead to crassness and incivility, and music to softness and frailty. "[But] he who best blends gymnastics with music and applies them most suitably to the soul," Plato wrote, "should most rightly pronounce to be the most perfect and harmonious musician, far rather than the one who brings the strings into unison with one another."[2] *Mens sana in corpore sano.* Plato believed music encompasses all that is most resistant to philosophy, making it the perfect complement to pure thought and science.[3] We ignore his guidance to our own detriment.

Despite musical education's current dereliction, it has enjoyed a long affiliation with mathematics and science. In the quadrivium, it was essentially a branch of mathematics. It focused more on musical intervals and harmonics than the playing of music. The great mathematician Gottfried Leibniz, who, along with Newton, invented calculus, tied musical structures to mathematics. To appreciate music he thought we must, at least at some level, also appreciate its mathematical framework. Philosopher Arthur Schopenhauer bristled at this idea. He argued that the exploitation of music as a mathematical tool was too mechanistic and utilitarian, completely missing its real essence:

> [In music we must] look for more than that "*exercitium arithmeticae occultum nescientis se numerae animi*" which Leibniz took it to be ... We must attribute to music a far more serious and profound significance that refers to the

innermost being of the world and of our own self.[4]

That is, while Leibniz asserted that "Music is a hidden exercise in arithmetic of a mind that doesn't know itself to be dealing with numbers,"[5] Schopenhauer framed it instead as "Music is a hidden exercise in metaphysics of a mind that doesn't know itself to be dealing with philosophy."[6] Rather than just responding to music as a stand-in for number, in other words, the mind is responding much more deeply to it.

In *The World as Will and Representation,* Schopenhauer claimed music is a universal language that expresses—*in mere tones*—the most distinct manifestations of inner beauty. He called this the aesthetic way of knowing, and it was a window into the essential nature of things. It bypasses mere ideas to speak to us directly. Echoing this book's epigraph by the poet Rainer Maria Rilke, Schopenhauer wrote:

> Music is by no means like the other arts, namely a copy of the ideas, but a copy of the will itself, the objectivity of which are the ideas. For this reason, the effect of music is so very much more powerful and penetrating than is that of the other arts, for these others speak only of the shadow, but music of the essence.[7]

Not that he placed all this wisdom exclusively in the hands of the composer. The composer, Schopenhauer believed, was merely a naïve messenger whose music communicated "the profoundest wisdom in a language that his reasoning faculty does not understand."[8]

FIGURE 10.

Schopenhauer's young protégé Friedrich Nietzsche was even more enamored with music. Though the two later grew apart, they both believed in music's power to inspire wisdom. To Nietzsche, raised as a musician, music alone made life worthwhile. By the age of fourteen, he already had forty-six musical compositions to his name. He attended a competitive music school and was an accomplished pianist.[9] The jury is still out on whether his compositions were any good. Some, including Franz Liszt, liked his music. His friend and composer Richard Wagner, on the other hand, had to excuse himself the first time he heard one of Nietzsche's compositions—because he was crying with laughter.[10] The German conductor Hans von Bülow called another of his pieces "the most undelightful and the most anti-musical draft on musical paper that I have faced in a long time."[11]

Nietzsche deemed music essential for many reasons. Primarily, it was a positive influence on life and the will. He wrote, "God has given us music so that *first of all* it might lead us upward."[12] It also served him on a purely emotional level. As a young man, at the age of fourteen, he wrote, "One has to consider all those who despise [music] as soulless creatures similar to animals."[13] Later, he added in *Twilight of the Idols*, "Without music, life would be a mistake."[14] Other times, music was a source of inspiration for Nietzsche. It stimulated ingenuity and helped him make connections. He was frequently seen attending concerts, notebook in hand, explaining, "Bizet makes me a better philosopher."[15] Music penetrated the deepest recesses of his mind, exhibiting incredible staying power. Even at the end of Nietzsche's life, mind wracked with the insanity of tertiary syphilis[vi] and barely able to speak, he could still capably play the piano.[16]

A vast retinue of philosophers and thinkers regard music as vital to our mental well-being. We now understand that music has a strong association with intelligence. This is not to be confused with the well-known Mozart effect—the idea that simply listening to classical music somehow improves the brain. The study behind the Mozart effect was decidedly flawed by "an artifact of presence."[17] Once the control arm's silence was replaced with a placebo, the advantage of listening to music disappeared.[18] Just like nearly everything else that is worth doing in life, there is no passive path to excellence.

Once we consider the neurobiology of music and intelligence, we can understand why subsequent studies failed to produce similar benefits. Music's influence comes from long-term, deliberate interaction. How long? In one study, children showed improvements in verbal intelligence with as little as twenty days of music training.[19] Another study found greater IQ improvements in students after thirty-six weeks of training.[20] Structurally visible changes in the brain can be demonstrated on an MRI after just a year of playing the violin.[21]

Fortunately for the rest of us, the benefits of music are not limited only to the young. For instance, one study found that

---

[vi] The ultimate source of Nietzsche's mental demise is still debated. It was likely neurosyphilis but may have also been due to manic-depressive psychosis, brain tumor, or multi-infarct dementia.

six months of piano lessons improved working memory and executive function in older adults.[22] Maybe you can teach an old dog new tricks if you sing the instructions. These studies are a remarkable demonstration of the brain's plasticity and, thankfully, they contradict Spanish neuroscientist Santiago Ramón y Cajal's "harsh decree" that adults cannot grow new neurons. "Everything may die," he wrote, but "nothing may be regenerated."[23] Mounting evidence suggests he was mistaken.

People frequently wonder how we can separate the effects of an enriched childhood from the music itself. Could we be conflating causation and correlation? Children who study music generally come from higher socioeconomic backgrounds, which could be a confounding factor. A study of Swedish twins sought to clarify this issue. They identified elderly twins who were discordant for music: that is, one played music, and the other did not. They were otherwise genetically and environmentally identical. Could something as simple as learning a musical instrument impact cognitive function over time? The results were stunning: the musical twin was 64 percent less likely to develop dementia.[24] This far outweighs the benefits seen by pharmacologic treatments for dementia. Drugs like Aricept (donepezil) and Namenda (memantine) attempt to treat Alzheimer's disease but have produced disappointing results. These drugs *might* slow the progression to nursing home placement by about six months on average. They have shown minimal impact on objective measures like formal cognitive assessments, where they average about one point of improvement on a thirty-point scale.[25] While we must be cautious about overinterpreting the results of any single study, the data suggest a real and meaningful change with long-term exposure to music, something that Big Pharma can only dream of doing.

Music builds greater connectivity between the two hemispheres. It's like adding a backup drive or doubling your RAM. This redundancy of functional brain tissue, beyond what is needed for daily living, is called cognitive reserve. Investigations like the famous nun study suggest that those with greater cognitive reserve may avoid displaying outward signs and symptoms of Alzheimer's disease, even when later autopsy demonstrates advanced disease.[26] Other researchers have observed that musicians have younger-appearing brains than their age-matched cohorts.[27] It appears that music, by improving the brain's plasticity and cognitive reserve, enables it to better absorb the effects of aging and injury.

The anterior cingulate cortex (ACC) plays a role in maintaining attention, and also plays a role in long-term memory of music. Intriguingly, it is largely spared by the degeneration inflicted by Alzheimer's disease.[28] This may be one reason music can endure the ravages of aging. It likely explains Nietzsche's ability to still play the piano long after he lost the ability to speak. A friend's mother-in-law suffers from advanced Alzheimer's disease. She still plays the piano from memory—even though she could not tell you what month it is. These anecdotes, if nothing else, speak to the power of music, which persists in the hidden crevices of the brain even after all other aspects of the self are long forgotten.

Figure 11.

The benefits of music on the brain are probably more prophylactic than therapeutic. Building neural connections and brain plasticity mitigate aging's effects on the brain. It is much more challenging to rebuild the factory after it has burned down than to put in the fire retardant ahead of time. However, music may also have a therapeutic role. For instance, a study of thirty-eight stroke patients found that the group given six to eight weeks of music therapy saw more improvement than those receiving the usual care.[29] In particular, they showed improvements in nondominant hand function and nonverbal communication.[vii] They also experienced less anxiety and depression, common complications after a stroke. Another study taught piano to thirty-one elderly patients with mild cognitive impairment—a precursor to dementia—who had never played music before. After six months, these patients showed significant improvements in cognitive function.[30] While there is still a long way to go in the study of who, what, when, where, and how to utilize music therapy most effectively, these early studies suggest that it will likely play a growing role in neurological rehabilitation programs in the future.

Businessman Charlie Munger, Warren Buffet's partner at Berkshire Hathaway, once claimed, "I did not succeed in life by intelligence, I succeeded because I have a long attention span."[31] Is attention span something that can be learned, or are we just born that way? Once again, the science of music sheds light on this question. It may, in fact, be one of music's most important contributions to intelligence. Simply put, it is impossible to play classical music without concentration, focus, and discipline. A tolerance for delayed gratification is invaluable

---

[vii] Throughout this book, for simplicity's sake, we will assume the most common anatomical layout of the brain. Almost all right-handers and more than 60 percent of left-handers have left-brain dominance. That leaves a mere 4 percent right-brain dominant. Complicating things further, true left-handers are not just mirror images of a right-hander, as is commonly supposed.

in music and in life. As one writer quipped, "You can buy an expensive violin, but you can't buy ten years of practice."[32]

Ten years is about the equivalent of ten thousand hours at three hours per day. Mozart may have begun writing music at age four, but his compositions were not *masterful* until well after he had put in ten thousand hours. This concept was introduced to the public in Malcolm Gladwell's book *Outliers: The Story of Success*.[33] In a debate with writer David Epstein, Gladwell half-jokingly argued that "Mozart composes at a very young age, but nothing—*it's terrible*—until he's in his [early- to mid-twenties]."[34] To say that this idea has been controversial would be an understatement. One scientific review lambasted it as fallacious: "The evidence is quite clear that some people do reach an elite level of performance without copious practice, while other people fail to do so despite copious practice."[35] This is probably true for many types of skills. However, at least regarding psychomotor and intellectual tasks, Gladwell's case appears reasonably strong. A study of seventy-six classical music composers found that all but three took at least ten years to write their first masterpieces.[36] Even the three exceptions were not far off: Paganini wrote the *Caprices* in his ninth year, Shostakovich's Symphony No. 1 also came in his ninth year, and Erik Satie's *Trois Gymnopédies* came in his eighth. It thus seems reasonable to say that, at least for music, adequate deliberate practice is necessary, but not sufficient, to achieve mastery.

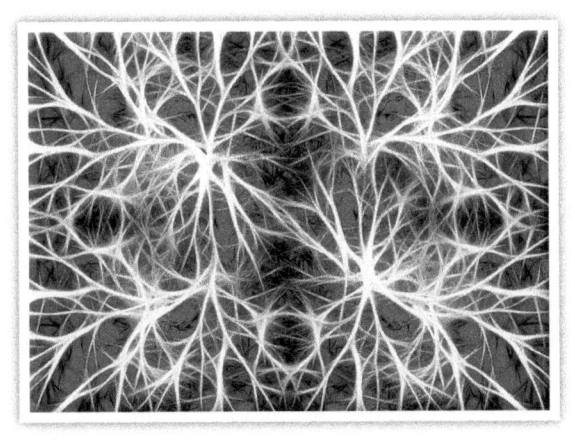

FIGURE 12: THE MESH OF CONNECTIVITY.

It is the mesh of connectivity between neurons, as opposed to the neurons themselves, that drives creative power. The same can be said about the connections between ideas. Einstein called this combinatory play, highlighting the tripartite dance between knowledge, imagination, and curiosity that takes place when we combine ideas from different fields. He considered it an essential feature of productive thought.[37] Newton's gravity, for instance, combined Galileo's parabolas with the ellipses of Kepler. Maxwell, of course, combined electricity and magnetism. And Einstein's relativity further combined these theories of Newton and Maxwell, resolving the conflict between electromagnetism and mechanics. Steve Jobs described how important combinatory play is for innovation:

> Creativity is just connecting things. When you ask creative people how they did something, they feel a little guilty because they didn't really do it, they just saw something. It seemed obvious to them after a while. That's because they were able to connect experiences they've had and synthesize new things.[38]

Author and journalist Arthur Koestler coined the term *bisociation* to describe this phenomenon. He wanted to make clear the distinction between "routine skills of thinking on a single plane, as it were, and the creative act, which ... always operates on more than one plane."[39] Bisociation differs from association in that it is combinatorial, creative, and right-brained. Associative thought, on the other hand, stays within a single matrix—logical and left-brained.

Despite Einstein's brilliance, his brain was no larger than yours or mine. When pathologist Thomas Harvey measured it after his death, it weighed a below average 2.7 pounds.[40] It's disappointing size puzzled scientists, just as mathematician Carl Friedrich Gauss's had years before.[viii] They were convinced that size mattered. This simple phrenology seemed so undeniable. After Einstein's death from a ruptured aortic aneurysm, Harvey absconded with his brain before the rest of his body went into the crematorium. Einstein's family, who did not give their consent for this act, was understandably not happy about the brain's situation. They feared it would become a pop-culture circus curiosity. Harvey refused to turn over the brain and was eventually fired from Princeton Hospital for the act.

Over the ensuing years, Harvey parceled out bits and pieces of the brain to various researchers.[41] These studies, in the end, did clarify a few things. Einstein's cortex—the gray matter consisting of the neurons themselves—was unremarkable. It was, if anything, smaller than average. One important finding, initially underappreciated, related to his white matter. Einstein's was quite a bit larger and denser than expected.[42] He had an impressive 73 percent more glial cells than average. When Dr. Marian Diamond presented these findings at a conference, the

---

[viii] An average male brain, for comparison, weighs 2.9 pounds. More on Gauss later, but his brain, for comparison, weighed 3.2 pounds. Though slightly above average, it still puzzled scientists of the day who expected much more. They resorted to measuring surface area in an attempt to better reflect its impressive capacity. (Dunnington, "Sesquicentennial," 413.)

audience found them "comical" and meaningless.[43] Unwittingly, their minds were closed to the idea that anything other than neurons could reflect intelligence. We now recognize that it's the glial cells, not the neurons, that best demonstrate genius. The utility is in the roads, not the factories. A brilliant mind is a good networker, with many interweaving attachments between the neurons. Connections between nerve cells are called synapses. The more connections, the more power.

The brain is often compared to a computer, but this is a not a perfect analogy. Computers are serial processors—they cannot do two things at once, even though their speed might make it appear otherwise. The brain, however, is a parallel processor. The average human brain has eight-six billion neurons, twelve times the number of people on Earth.[44] Each one is powerfully connected to hundreds, or even thousands, of other neurons. These neural networks dynamically manage multiple simultaneous processes, yielding exponentially more power than is possible with a computer's serial processor.[45] To put this in perspective, Nobel Prize-winning neuroscientist Gerald Edelman pointed out that it would take more than thirty-two million years to count all the synapses in the human brain at one synapse per second. If we counted all the connections, it would take an unimaginable amount of time: ten followed by a million zeros.[46] The computing power of our brain outstrips even the most powerful modern supercomputers.

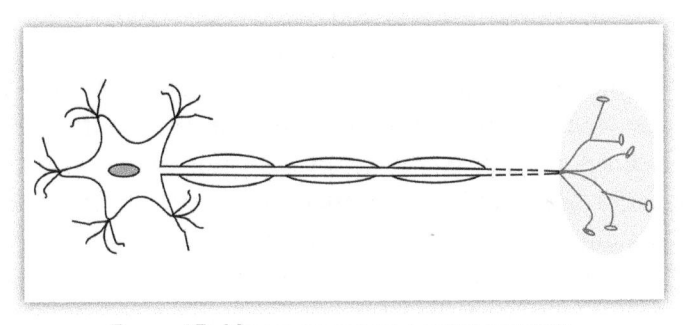

FIGURE 13: MYELIN SHEATH ON A NERVE DENDRITE.

Daniel Coyle reviewed the concept of interconnectivity in his book *The Talent Code*. "Skill is myelin insulation [figure 13] that wraps neural circuits and grows according to certain signals," he wrote.[47] And just like wire insulation, it boosts the speed of the electrical signal. The analogy of brain circuits to electrical circuits is remarkably close to reality. Skill and ability are lost when diseases like multiple sclerosis start unraveling this insulation. Myelin and its supporting glial cells—like oligodendrocytes and astrocytes—make up more than half of the brain. This is the white matter that impressed Dr. Diamond. Studies show a strong correlation between the amount of white matter and the hours of practice in pianists.[48] Every skill is essentially a memory. Eventually, with ongoing practice, it becomes hardwired into a vast myelinated web.[49] As Coyle noted, prewiring for skills would be evolutionarily foolish. It would be rigidly inflexible. Adaptability is the opposite—it is robust, even antifragile. Philosopher Nassim Taleb described antifragility as something that doesn't just tolerate stress (robustness) but improves with it.[50] Adaptability bestows an ability to learn *any* skill, not just the ones we are born with. The ten-thousand-hours rule essentially recognizes the time it takes to build this myelin latticework. It is a dynamic interaction between necessity and extravagance. The incessant tension between training and detraining plays out on the scale of these neuronal connections. The web is only hardwired to the extent that it is maintained. The brain is an energy hog that burns nearly a quarter of our total caloric expenditure—despite weighing in at a mere 2 percent of our total weight. Evolution's relentless efficiency will not expend precious resources maintaining a connection for which there is no convincing need. Franz Liszt put this unremitting push and pull of skill acquisition in perspective when he said, "If I miss practicing one day, I know it; if I miss two days, my friends know it; and if I miss three days, the public knows it."[51] Natural selection constantly seeks the most efficient means to an end. And in the brain, it has crafted a powerfully adaptable machine.

A:

Descent with Modification

MORE THAN TWO THOUSAND YEARS AGO, ARISTOTLE WROTE, "THE BRAIN IS bipartite in all animals."[1] Humans are unique, however, in possessing true hemispheric specialization. While our primate relatives do possess some left-right specialization, it is absent in almost all other mammals.[2] In the human embryo, the right hemisphere develops first. The left brain does not become dominant until much later. According to recapitulation theory, this suggests the right brain is more primitive. This theory was formulated in the 1790s by Johann Friedrich Meckel. It states that the development of an embryo mimics the evolutionary past of that species. Ernst Haeckel later expanded recapitulation theory in an attempt to unite Lamarckism and Darwinism. His maxim, "ontogeny recapitulates phylogeny," reflects these now outmoded beliefs. Modern scientists find many problems with recapitulation theory but still recognize that embryogenesis can subtly reflect evolutionary history.[3]

The vast extent over which music is spread in the brain should hint at its importance in hominid evolution. Indeed, the brain needs to maintain numerous disparate elements independently to interpret even just the basic components of music: pitch, rhythm, melody, harmony, tempo, meter, duration, and loudness. Furthermore, all these components are spread, like jam on toast, throughout the brain. They are not concentrated into any such "nucleus musicorum" or "pars lyrici." It is as if nature has squirreled them away to ensure against any damage affecting our ability to interact with music. Much the same as you might diversify your retirement portfolio, the brain doesn't put all its musical eggs in one basket. Damage to the left brain, for example, affects rhythm, but meter, on the right side, remains intact. The cerebellum is the timekeeper. The mesolimbic system—the reward system—ensures we remember music. Songs from childhood come back with ease, though we struggle to recall the memorized-five-times-already preamble to the US Constitution.

FIGURE 14.

Scientist and naturalist Edward O. Wilson highlighted the relationship between biology and culture when he wrote: "History makes little sense without prehistory, and prehistory makes little sense without biology. Knowledge of prehistory and biology is increasing rapidly, bringing into focus how humanity originated and why a species like our own exists on this planet."[4]

Music is both biological and cultural. Cultural evolution occurs much more quickly than biological evolution. Once humankind adopted music, it took off. More than fifteen hundred years ago, the Roman philosopher Boethius already appreciated music's deep evolutionary roots: "Music is so naturally united with us," he wrote, "that we cannot be free from it even if we so desired."[5]

Like any highly successful trait, music is now ubiquitous. "Music is part of being human," wrote Oliver Sacks, "and there is no human culture in which it is not highly developed and esteemed."[6] Numerous prehistorical artifacts bespeak its ancient pedigree. Cave paintings depict dancers and flautists. A forty-three-thousand-year-old flute, carved from the femur of an extinct European bear, was found in a Slovenian cave.[7] This was at first believed to be of Neanderthal making. However, scientists now feel that this ersatz flute is just an artifact, the holes likely made by hyena teeth.[8] Nonetheless, genuine *Homo sapiens* flutes made of naturally hollow vulture bones date back more than thirty-five thousand years.[9]

In either case, it is evident that music predated agriculture. It even, more than likely, predated language. That's right, our early ancestors likely sang and danced several hundred thousand years before we spoke.[10] This is not a novel idea. Jean-Jacques Rousseau, an accomplished composer in his own right, described primitive languages as "melodic and poetic, rather than practical or prosaic."[11] Charles Darwin likened us to the gibbon, whose musical singing spans an octave by half tones—just like our chromatic scale. This and other examples led him to infer, "The progenitors of man probably uttered musical tones before they had acquired the power of articulate speech."[12] Even Sherlock Holmes weighed in on the subject:

> "Do you remember what Darwin says about music? He claims that the power of producing

and appreciating it existed among the human race long before the power of speech was arrived at. Perhaps that is why we are so subtly influenced by it. There are vague memories in our souls of those misty centuries when the world was in its childhood."[13]

The Neanderthals did not go extinct until about forty thousand years ago. It is likely that *Homo sapiens* and Neanderthals interacted and crossed paths for many thousands of years. Whether these two hominids influenced each other's culture is unknowable. We do know that *Homo sapiens* and Neanderthals both share similar auditory and speech capabilities.[14] Archeology professor Steven Mithen suggested that the Neanderthals may have possessed a "musilanguage" that was concomitantly proto-music and proto-language.[15] Yuval Harari asserted in his book *Sapiens* that our ancestors first developed nuanced language about seventy thousand years ago. It is likely our own, more advanced language skills—and the cooperation and group dynamics it enabled—eventually led us to dominate our hominid competitors.[16]

Ethnomusicologist Joseph Jordania recently proposed that music may have had an entirely different teleology.[17] Humans, he observed, are the only ground-dwelling animal that sings. Other singing animals fly (birds), swim (whales), or live in the trees (gibbons). As a result, they are not as vulnerable if their singing is heard by a potential predator. In contrast, our ancestors may have explicitly exploited music to make themselves *more* obvious. This type of conspicuous marketing is known as *aposematism*. It is the polar opposite of camouflage. Compare it to the bright colors of a poison dart frog that caution potential predators not to eat them. Our ancestors originally came down from the trees as scavengers, not hunters. Troops of early humans followed prides of lions to steal their food.

How did hairless, clawless, and fangless humans accomplish this intimidating task? With music! Jordania argues that singing may have evolved to facilitate *confrontational scavenging*. These displays can be powerfully intimidating: watch the New Zealand rugby team's haka,[ix] and you'll get the picture. Add sticks and stones to the mix, and it is easy to see why the more powerful lions might have turned tail and scampered away.

Early languages, evolving out of this musical background, were likely melodic and poetic in nature. The descendants of these early tonal languages are still in use today. They include such complex and beautiful languages as Navajo and Mandarin. Tonal languages rely on pitch to convey meaning. The close relationship between music and tonal languages may explain why these ethnicities produce so many top musicians. It has been found that native speakers of tonal languages are much more likely to have perfect pitch. Among those who began musical training before the age of five, only 14 percent of American music students have perfect pitch. In Chinese students, this number rises to an extraordinary 60 percent![18] Noam Chomsky suspected that humans have the innate capacity to learn any language at birth; the brain is pruned and shaped by early experience.[19] Music leads to language and language back to music. The two are perpetually intertwined like strands of DNA.

Music's relationship with communication is more than just intriguing evolutionary trivia. Music proves more effective than prose for memorization. Consider the earworm: the song that gets stuck in your head for hours despite your best efforts to banish it. This is the mnemonic power of music. It is why epic poems were sung in verse. At the extreme end of the spectrum are people with a phonographic memory. Like those with a photographic memory, these people can remember even complex music as if it were stamped in their brains. Oliver

---

[ix] For an intimidating performance of the haka: https://www.youtube.com/watch?v=yiKFYTFJ_kw.

Sacks described one man who knew more than two thousand operas by memory.[20] This is obviously an extreme example, but all of us, even without this special ability, have a greater capacity for remembering music than we do for unadorned speech.

Australian Aborigines refer to music as a "memory bank for finding one's way about the world."[21] Just like a bee's waggle dance communicates directions to other bees in the hive, music creates a latticework within the mind. Spatial patterning and its relationship to memory is highlighted by a study of chess grandmasters. These elite players memorize complex board configurations vastly more effectively than the average person. However, if the pieces are placed at random—not in an authentic game's pattern—then the grandmasters do not remember any better than the average woodpusher.[22] A solo violinist performing a long concerto by memory faces similar issues. He or she needs a framework by which to play. These patterns become so entrained that something as simple as a bow misdirection can throw the whole thing off the rails. At this point, the violinist must either skip ahead to the next section or restart from the beginning.

Jascha Heifetz was a legend in the early recording days of classical music. The *New York Times* called him "a virtuoso of frightening perfection."[23] Even he had his moments. One evening in Dallas in 1954, he got off track near the beginning of the third movement of the Sibelius Violin Concerto. He signaled to the conductor, Walter Hendl, to start over, and they completed the concerto beautifully.[24] The next morning headlines around the country decried his "fluff" and obsessed over how "the perfectionist forgot today."[25] Heifetz and Hendl were surprised by all the media attention. They had not considered it anything more than a minor nuisance.

FIGURE 15: JASCHA HEIFETZ AS A YOUNG MAN.

E:
Epiphany

MUSIC'S EARLY CONNECTION WITH MATHEMATICS GREW OVER TIME INTO A connection with scientific thought as well. By the time of Einstein and his fellow musician physicists, it had become so entwined with quantum mechanics that it was impossible to consider one without the other. One of the fathers of quantum mechanics, Arnold Sommerfeld, was a dedicated pianist.[1] He noted the profound musicality of the Niels Bohr and Max Planck atomic models:

> The language of spectra is a true "music of the spheres" in order and harmony that becomes ever more perfect in spite of the manifold variety. The theory of spectral lines will bear the name of Bohr for all time. But yet another name will be permanently associated with it, that of Planck. All integral laws of spectral lines and of atomic theory spring originally from the quantum theory. It is the mysterious organon on which Nature plays her music of the spectra, and

according to the rhythm of which she regulates the structure of the atoms and nuclei.[2]

The physicist Louis de Broglie drew a further analogy between the discrete set of energy levels that characterized the atom and the set of mechanical vibrations observed in violin strings.[3] In quantum mechanics, a quantum leap occurs when an electron absorbs or emits enough energy to jump into another orbit—like the steppingstones of a musical scale. Insight, like a quantum leap, is not gradual or linear. It is an inspired jump to another level.

Creativity is the "rambling and volatile power of imagination," according to William Duff.[4] Two hundred years before Einstein, he defined genius as the outward expression of combinatorial wisdom:

> Imagination is that faculty whereby the mind not only reflects on its own operations, but which assembles the various ideas ... treasured up in the repository of the memory, compounding or disjoining them at pleasure; and which, by its plastic power of inventing new associations of ideas, and of combining them with infinite variety, is enabled to present a creation of its own ... which never existed in nature.[5]

I was struck by his use of the term *plastic*, which predates the existence of actual plastic substances. Plastic, in the material sense, wasn't in use until the early 1900s. Taken from the Greek *plastikos*, it refers to something that is capable of being molded or shaped. Plastic, in this sense, is perfectly descriptive of creative genius. The creative mind molds and shapes itself for whatever new idea is needed. Thinking like Sherlock Holmes, outside the box, requires plasticity.

A creative approach to *all* problems would be burdensome.

Like trying to herd cats, insight is capricious, unpredictable, and hard to control. It is analogous to Daniel Kahneman's type 2 thinking.[6] Type 1 thinking—efficient, intuitive, and quick—is better for the quotidian decisions that are best made on autopilot. It would be exhausting to consciously think through everything we do on a daily basis. Important decisions with large repercussions, however, should be more robustly thoughtful to minimize cognitive biases. The same goes for creativity. Most of our thinking can, and should, be logical. But when we are puzzled, it pays to get creative. It is time to get off track and bring in the right brain. This wandering and explorational aspect of creativity is highlighted by the humorous and insightful Persian poet Hafiz:

> The mule I sit on while I recite
> Starts off in one direction
> But then gets drunk
> And lost in
> Heaven.

The usual path is well trammeled. There is nothing new to find there. "Improvement makes strait roads," wrote William Blake, "but the crooked roads without improvement are roads of genius."[7] Physicist David Bohm emphasized the importance of taking the road less traveled in science. "One prerequisite for originality," he wrote, "is clearly that a person shall not be inclined to impose his preconceptions on the fact as he sees it."[8] Even the great Einstein could get stuck, sometimes for years. Once, when he was baffled by Mercury's orbit, Einstein despaired, "I do not believe that I myself am in the position to find the error, because my mind follows the same old rut too much in this matter."[9] He did eventually get out of that rut, but it required his entire creative toolbox to do so.

Divergent thinking like this is not easily summoned. It is part and parcel of finding that faint path wending through the willows. Some of them are just dead ends. Curiosity is what drives us to keep looking for a way through. "I have no special talents," Einstein once volunteered, "I am only passionately curious."[10] His hero, Michael Faraday, similarly wrote, "Do not suppose that I was a very deep thinker, or was marked as a precocious person. I was a very lively imaginative person, and could believe in the *Arabian Nights* as easily as in the *Encyclopaedia*."[11]

Creativity does not closely mirror standard intelligence. The average IQ of highly creative people is around 120, comparable to schoolteachers, nurses, and managers.[12] High-level scientists average closer to 140. This trivial fact only serves to emphasize that IQ is a poor measure of original thought. Indeed, there is some evidence that creativity may even be inversely proportional to IQ. Those with higher IQs may find themselves more stuck in a rut, as noted by writer Adam Grant:

> Mental horsepower doesn't guarantee mental dexterity. No matter how much brainpower you have, if you lack the motivation to change your mind, you'll miss many occasions to think again. Research reveals that the higher you score on an IQ test, the more likely you are to fall for stereotypes, because you're faster at recognizing patterns. And recent experiments suggest that the smarter you are, the more you might struggle to update your beliefs.[13]

Physicist Richard Feynman, in particular, epitomizes the problem with IQ scores. His was a merely respectable 125.[14] And yet Feynman was universally considered brilliant. Heck, *Genius* is even the title of his biography.

Many bright scientists work to change the world for the better. "But Einstein," science writer James Gleick noted, "seemed to have reinvented it whole."[15] Mathematician Mark

Kac, attempting to distinguish between smart and genius, made the following comparison:

> There are two kinds of geniuses: the "ordinary" and the "magicians." An ordinary genius is a [person] that you and I would be just as good as, if we were only many times better. There is no mystery as to how his mind works. Once we understand what he has done, we feel certain that we, too, could have done it. It is different with the magicians ... Even after we understand what they have done, the process by which they have done it is completely dark.[16]

Kac placed Feynman in the latter category: a true magician. Magicians and child prodigies have much in common. They seem to be disproportionately drawn to highly abstract fields like music, mathematics, and chess. Musician, mathematician, and writer Edward Rothstein noted that these fields are less dependent on empirical experience than with insight into a seemingly closed and abstract universe.[17] Gauss was one such mathematical prodigy. He was allegedly correcting his father's mathematical calculations by the age of three. The musical prodigy Felix Mendelssohn had already composed twelve symphonies by the age of seventeen. Music and math are so abstract that they concurrently seem both otherworldly and this-worldly. Yet, says Rothstein, "There is something more than abstract about mathematics and music."[18] They also have the power to transform the world.

When it came to music, Feynman was no Einstein. He did not consider himself the least bit musical. He was, however, an unabashed bongo player—something he picked up while on sabbatical in Rio de Janeiro.[19] He claimed to be tone-deaf, insisting that melody and harmony were like "sand in the mouth."[20] He found classical music boring. Once, on a bet, he tried to teach himself Nikolai Rimsky-Korsakov's *Flight of the*

Bumblebee.[21] He failed miserably. Nonetheless, Feynman grew to love improvisation and rhythm. In homage to Galileo, Feynman and his Princeton colleague John Tukey investigated their ability to keep track of time by counting. They found that different brains keep track of rhythm in different ways. Feynman's was aural, for instance, whereas Tukey's was visual.[22]

Plato described rhythm as a key component of music, given by the muses to help us prevail over our "irregular and graceless ways."[23] Feynman did not lack for grace. He adeptly played two-handed polyrhythms: "Not just the usual three against two and four against three but—astonishing to classically trained musicians—seven against six and thirteen against twelve."[24] He had found his musical outlet. Composer John Cage experimented with similar concepts. At times, he dispensed with pitch altogether and instead focused attention on rhythm and timbre. He too experimented with very challenging polyrhythms, as in his *Third Construction*.

Edith Sitwell's "flamboyant diction and cacophonous, jazzy rhythms" won over Feynman's heart, and she quickly became one of his favorite poets.[25] She compared sound's reliance on rhythm to sight's reliance on light. Feynman could easily sense the intrinsic beat, and he exuberantly recited her poetry. Like Feynman's polyrhythms, the universe doesn't settle for just one beat. "At the minute scale of the grains of space," wrote Carlo Rovelli, "the dance of nature does not take place to the rhythm of the baton of a single orchestral conductor, at a single tempo: every process dances independently with its neighbors, to its own rhythm."[26]

Mathematician Tobias Dantzig, in *Number: The Language of Science*, described the eternal issue of how mathematics deals with time. Music solves this problem. The way it structures time is particularly relevant to physics. Thus music, wrote German philosopher Georg Wilhelm Friedrich Hegel, "is the first art that gives us the sense of unification occurring in time rather than in space."[27] That explains its placement as a pillar of the quadrivium, and it is one of music's most essential functions.

In general relativity, both space and time bend toward matter. Arthur Stanley Eddington proved Einstein's theory when he showed that starlight bends as it passes the sun. The sun is the only nearby object with enough mass to definitively demonstrate this phenomenon, but it is usually too bright to see the stars beyond it. The solar eclipse expected on May 29, 1919 provided a rare opportunity to test the theory's predictions. Months in planning, the team traveled to Spanish Guinea off the west coast of Africa. Though hampered by a terribly timed thunderstorm, the skies cleared at just the right moment. "It *was* a miracle," wrote one of the expedition members, "that little lake of blue sky in the center of which was the phenomenon we had all gathered to see."[28] Moments later, the area clouded over again. Eddington took enough photos during that brief clearing to prove that the stars appeared closer together after their light passed by the sun. Time, like light, should theoretically do the same thing. If only we could see it.

Perhaps this obsession with time explains Einstein's extensive use of music as he pondered general relativity and its implications for the space-time continuum. He believed, for instance, that physical time could not be precisely defined.[29] This may have just been his excuse for missing several entrances while playing with the famed pianist Artur Schnabel.[30] Another time he got out of sync with the legendary violin virtuoso Fritz Kreisler.[31] Both musicians exasperatingly ribbed him, expressing wonder as to whether the professor knew how to count. Most of the time, though, Einstein found the time structures of music intuitive. He even bragged to his childhood duet partner, Hans Wohlwend, that he did not need to count the beats because they were in his blood.[32]

The ultimate sources of creativity remain a mystery. Even highly creative people are often unaware of where their ideas come from. John Cage said that new ideas can sometimes just "fly into one's head like birds."[33] Leonard Bernstein tried to explain:

> The mind, where all this creativity takes place, is an immensely complicated circuitry of electronic threads, all of which are connected at a certain point and are informational. But every once in a while, there is something like a short circuit; two of them will cross, touch, and set off something called an idea. This is the most exciting moment that can happen in an artist's life.[34]

But these comments do not explain *how* to arrive at such a point. Creativity is mercurial; it does not just come on command. To account for its capricious nature, Steven Pressfield invoked *the muse*: creative instinct, divine inspiration.[35] This idea originated with Hesiod and Homer, who also prayed to the muses for inspiration. Parmenides relied on Dikē, the guardian of truth. Regardless of how one tries to conjure creativity, the fact of the matter is, it's hard! Music, though, can summon the muse better than anything else in the creativity toolbox.

FIGURE 16.

Sometimes the muse produces pure serendipity—*eureka!* —just like Einstein's "I've got it!"[36] The story may be apocryphal,[x] but Archimedes is said to have first muttered, "Eureka!" when he saw the rising level of bath water as his body sank into the tub.[37] One characteristic feature of insight is how frequently it appears obvious in hindsight. Like John Cage's joke, in response to comments about his experimental music: "If music's that easy to write, [anyone] could do it." To which Cage responded, "Of course they could, but they don't."[38] Breaking rules you don't know is not creativity. One must, like Cage, have a deep understanding of the system he or she is trying to subvert if it is to be meaningful.

Astronomer William Morgan observed, "So much of the creative process takes place in the subconscious ... It is in such uncharted areas of mental space where some of the deepest science has its origins."[39] The muse symbolizes the difficulty we face in attempting to access this uncharted realm. The muse's hideout might not remain uncharted for long, though. It is probably in the right anterior superior temporal gyrus. This fold of brain tissue explodes in a burst of electrical activity just before a moment of insight.[40] In the same way that you look up or close your eyes when you're trying to recall something, there is a brain blink just prior to this burst of activity. On EEG tracings, a brain blink consists of slow sinusoidal alpha waves.[41] The mind is trying to reduce distractions from other parts of the brain: "Shhhh! Quiet! I'm trying to think!" The fireworks, in contrast, are high-frequency gamma waves. Researchers don't see this pattern of activation during logical and analytical thought—they only see it at the moment of insight.

---

[x] Archimedes was asked by King Hiero II of Syracuse to determine whether his crown was solid gold. How to do so without damaging the crown? When he saw the bath water rise, he instantly knew that he could measure the volume of the crown and compare its density with that of solid gold. As it turns out, the goldsmith had substituted a cheaper alloy, pocketing the excess gold for himself.

Intriguingly, though probably not too surprising at this point in the story, the anterior superior temporal gyrus also plays a crucial role in processing music.[42] It lights up on functional MRI when music is played or heard.[43] Most people only activate the right side, but musicians, with their two hemispheres working as a team, light up both sides.[44] This insight center is closely connected with the anterior cingulate cortex (ACC), an area that directs attention and monitors for the presence of competing ideas. It can broaden or narrow the scope of possibilities you are prepared to consider. Like the anterior superior temporal gyrus, the ACC is suppressed by analytical thought and activated by a creative mindset.[45] Sometimes analytical thought is helpful. It induces mental tunnel vision, helping to focus on the task at hand. The ACC is versatile, though, and the converse is also true: when insight strikes, the ACC helps shut down logical thought so we can run wild with the new idea. The muse is disorderly and rebellious and wants you open to new ideas. Sounds like Einstein, doesn't it?

English social psychologist Graham Wallas described four stages of scientific insight.[46] The first is *preparation*: the hard work of mastering a subject. Einstein and Fleming sowed their seeds of inspiration in fertile soil. The next stage is *incubation*. Divergent thinking stews in the subconscious, percolating through the brain, sorting and sifting while sleeping, meditating, or playing music. Then, seemingly out of the blue, wires cross, a spark is created, and eureka! —*illumination*. Finally, there is *verification*. First to ourselves and then to others, we must prove that the idea really works. This final stage is a key difference between artistic creativity and scientific creativity. For this reason, Feynman felt that imagination in science is misunderstood and underappreciated. Whatever the scientist imagines, he told his students in a lecture on Maxwell's equations, must be entirely new and yet still consistent with everything else we know about reality. It is imagination "in a straitjacket, so to speak."[47]

It would be a tremendous asset to possess a shortcut to creativity. Music is this tool. That is why Einstein, Holmes, and countless others made it such an integral part of their routines. Author Ursula Le Guin, whose daughter Elisabeth is a cellist, wrote about her own discovery of the overlap between music and creative thought. "It had never occurred to me that music and thinking are so much alike," she wrote. "In fact, you could say music is another way of thinking, or maybe thinking is another kind of music."[48] Music prepares the mind for complex pattern recognition and language processing. It coaxes us into the more creative right brain, which gets us out of the box and off the beaten path. It opens the door to serendipity and clears the way for finding connections that the logical left brain has missed. This is literally divergent thinking, and it is a prerequisite for creativity and originality. It is seeing the water rise in the tub, noticing it, and then seeing how it relates to another problem.

B:
Disorder and Distraction

Yehudi Menuhin was one of the great violinists of the twentieth century. Einstein was lucky enough to attend his debut with the Berlin Philharmonic in 1929. Menuhin, only twelve years old, played concertos by Bach, Beethoven, and Brahms. Einstein was so moved by Menuhin's performance that he rushed to meet him after the show, exclaiming, as Menuhin recounted, "Now I know that there is a God in heaven!"[1]

Figure 17: A portrait of the young Yehudi Menuhin, painted shortly after Einstein saw him perform at the Berlin Philharmonic.

Menuhin shared many of Einstein's views on music. "Music creates order out of chaos," he wrote, "for rhythm imposes unanimity upon the divergent, melody imposes continuity upon the disjointed, and harmony imposes compatibility upon the incongruous."[2] This was consistent with what philosophers had recognized for millennia. Plato argued that music is heaven-sent to "correct any discord which may have arisen in the courses of the soul, and to be our ally in bringing her into harmony."[3] That is what Einstein glimpsed when he played Mozart: hidden universal order! It was fundamentally analogous to his quest for a unified field theory. In his *Act of Creation*, Koestler wrote, "If the explanations of science are like streams joining rivers, rivers moving towards the unifying ocean, the explanations of art may be compared to the tracing back of a ripple in the stream to its source in a distant mountain-spring."[4] The great physicist Freeman Dyson once had the fortune of hearing John Maynard Keynes speak at Cambridge University. Describing Isaac Newton, Keynes said that great geniuses see the universe as a riddle, with clues spread out by God in a sort of "philosopher's treasure hunt."[5] This struck a chord with Dyson. Einstein also endeavored to uncover enough of these clues to reveal a fundamental pattern. He tirelessly strove to trace that tiny ripple to its source. "Nature shows us only the tail of the lion," he wrote his friend Heinrich Zangger. "But there is no doubt in my mind that the lion belongs with it even if he cannot reveal himself to the eye all at once because of his huge dimensions. We see him only the way a louse that sits upon him would do."[6] Neuropsychologist Donald Berlyne depicted the common thread of genius—*in any field*—as an incessant striving toward unity. "The aim of all intellectual pursuits, including science, philosophy, and art," he wrote, "is to seek unity in the midst of diversity or order in the midst of complexity."[7] This described Einstein to a T.

Einstein and Bohr had many heated debates about quantum

mechanics. Einstein was irritated by what he considered its essential ugliness. It seemed to portray the world as less simple, less understandable, and less beautiful. Certainly, the world couldn't be *this strange.*[8] Eventually, Einstein admitted that quantum mechanics might be a useful tool, but this didn't stop him from spending the rest of his life trying to improve upon it. The influence of quantum mechanics persists today, and physicists still—one hundred years later—don't agree on what it means.[9] As the great British biologist J. B. S. Haldane observed in *Possible Worlds and Other Papers,* "The universe is not only queerer than we suppose, it is queerer than we *can* suppose."[10] It proves inestimably difficult to imagine something beyond our own experience.

It is undoubtedly naïve to envision the universe as a perfectly neat and tidy Eden. Nature, emphatically, requires disorder. The laws of thermodynamics insist upon it. Entropy, the second law of thermodynamics, defines the propensity for disorder to increase over time. It was the chief focus of Max Planck's early research. So, when he spoke with Einstein about unifying the forces of nature, he knew the challenges ahead of them. Neither evolution, nor life, could exist without entropy and disorder. But this does not mean that there is no pattern or beauty within the clutter. Music creates a perfect analogy for this aspect of nature: spotlighting shadowy patterns, symmetries, and connections that enlighten our understanding of the universe. Perfect order is boring! Music without surprise is insipid. Art historian Ernst Gombrich noted, "Delight lies somewhere between boredom and confusion."[11] Richard Powers in the novel *Orfeo* added, "Even beauty exhausts itself and leaves the ear wanting other sounds."[12]

By the early nineteenth century, chemists had dragged their nascent science out from under the mysticism of alchemy. They had broken away from the mistakes of the past, like Aristotle's four elements and phlogiston theory, but the field was still in chaos. Sixty-three elements were known. It was unclear how many more were still out there and what properties they might have.[13] Seeking order and pattern would be required if there was to be further progress in the young science. John Dalton had already established the concept of using atomic weights to identify elements in his book *A New System of Chemical Philosophy*, published in 1808. British chemist John Newlands thought he noticed a pattern in these numbers. When the known elements were placed in order of atomic mass, certain properties seemed to repeat every eighth element. Newlands, who had been trained by his Italian mother in music as a child, metaphorically called this the "Law of Octaves."[14] His peers laughed at this idea, joking that he might as well have just arranged the elements in alphabetical order.[15] But he was on the right track, and his idea fit even better once the noble gases were discovered.

The work of putting the elements in order was eventually completed by Dmitri Mendeleev, who had no knowledge of Newlands's previous attempt but used a very similar framework. The science of chemistry exemplifies the progress that can be made in a field once pattern and beauty are recognized. That is why these traits are so highly regarded. The periodic table's order led to the eventual discovery of many missing elements. It continues to inform chemists of similar properties within each column.

Music-inspired science relates just as easily to chemistry as it does to physics. The main character in Richard Powers's novel *Orfeo* was inspired to become a musician early in life by the beauty of Mozart's *Jupiter* symphony. When he later became a chemist, he noted the similarity between his two chosen fields: "The symmetries hidden in the columns of the periodic table had something of the *Jupiter's* grandeur."[16] Many real-world scientists have also integrated music and chemistry. Aleksandr Borodin was as a professor of chemistry at the Imperial Medical-Surgical Academy in St. Petersburg; composing was just his side hustle. And composer Sir Edward Elgar was also a serious chemist, even successfully patenting an apparatus that produced hydrogen sulfide gas. Thirty years ago, geneticist Susumo Ohno creatively converted bits of DNA into musical equivalents that are remarkably reminiscent of a Chopin nocturne.[17] Now you can upload your own DNA to a company in Edinburgh, Scotland that creates your own unique symphony based on your genetic sequences.[xi]

When Sherlock Holmes was hunting obscure patterns, he played the violin. In *The Adventure of the Norwood Builder*, Watson relates a story that highlights how Holmes used music to reveal hidden structure—just as Einstein did in his efforts to clarify the universe's murky harmonies:

> For an hour he droned away upon his violin, endeavouring to soothe his own ruffled spirits. At last he flung down the instrument and plunged into a detailed account of his misadventures. "It's all going wrong, Watson—all as wrong as it can go ... All my instincts are one way and all the facts are the other.

Playing music activated his pattern recognition software. It encouraged different parts of his mind to focus on a problem that logic alone was struggling to solve. Music has the capacity to get the mind off the multitasking treadmill. "I am a believer in unconscious celebration," wrote Alexander Graham Bell, himself a talented pianist. "The brain is working all the time, though we do not know it."[18] In this way, it starts to look a lot like meditation. Einstein never called himself a meditator, but the term is an apt description of his thought process. He

---

[xi] Your data from *Ancestry.com* or *23andMe* can be uploaded to *Your DNA Song*, at https://www.yourdnasong.com.

enjoyed allowing his unconscious to roam over hill and dale hunting for concealed connections. "I have no doubt that our thinking goes on ... to a considerable degree unconsciously," he wrote in his autobiography.[19] Ellen Langer's "illusion of control" tricks us into believing our conscious mind is the more potent one.[20] It reassures us to believe we have more power to control outcomes than we actually do. But it's the converse that is true: the unconscious mind is vastly more powerful. Our conscious mind can manage about forty pieces of information per second. This seems impressive—until compared to the unconscious mind's eleven million bits per second![21] But unless we can harness this power, it is only so much daydreaming.

FIGURE 18.

Buddhist meditators seeking enlightenment find it opens the mind to different ways of seeing things. Zen scholar D. T. Suzuki echoed Schopenhauer's description of music[xii] when he wrote, "The essence of Zen Buddhism consists in acquiring a new viewpoint on life and things generally."[22] Nowadays meditation is experiencing something of a renaissance. Still, most people

---

[xii] As quoted earlier: "... the effect of music is so very much more powerful and penetrating than is that of the other arts, for these others speak only of the shadow, but music of the essence."

think of it as a seated technique primarily used to improve introspection. But there are many roads to the meditative state, and many different reasons to do it. Lucy Lippard wrote about how the Inuit use walking as a form of meditation:

> An Eskimo custom offers an angry person release by walking the emotion out of his or her system in a straight line across the landscape; the point at which the anger is conquered is marked with a stick, bearing witness to the strength or length of the rage.[23]

The peripatetic philosophers, like Plato, Aristotle, and the Stoics, were so named because of their prodigious walking routines. Later, Jean-Jacques Rousseau, Henry David Thoreau, and Søren Kierkegaard carried on this great tradition. Gustav Mahler routinely took three-hour walks after lunch. Einstein walked nearly everywhere he went. According to his second wife, Elsa, he may not have had many options. "The professor does not drive," she divulged. "It's too complicated for him."[24] But it did provide him with plenty of thinking time. Frequently he had company on his walks. Colleagues wishing to discuss ideas often found a sympathetic ear if they joined him during his commute.

For Einstein, walking, sailing, and music were all akin to meditation. He loved cruising his sailboat around Lake Saranac or Long Island Sound. When the wind died, he would float aimlessly, all the while scribbling notes until it picked up again.[25] As with driving, he was also occasionally challenged by sailing. A *New York Times* headline mocked one mishap with the headline: "Relative Tide and Sand Bars Trap Einstein."[26] Nonetheless, it remained one of his favorite activities.

FIGURE 19: EINSTEIN SAILING NEAR RHODE ISLAND, CIRCA 1934.

Meditation reproduces many of the same neurologic patterns we see with music. A study of experienced meditators found increases in the cortical thickness in regions related to motor, sensory, auditory, visual, and interoceptive processing.[27] This suggests that meditation may improve the functional plasticity of the brain in ways similar to music. A follow-up study sought to determine whether meditation could do the same thing in beginning meditators. Eight weeks of training did, remarkably, induce visible changes on MRI.[28] The findings were particularly prominent in the hippocampus, an area implicated in many emotional disorders. In a vicious cycle, stressful experiences diminish the size of the hippocampus, while a small hippocampus predicts further susceptibility to stress-related disorders.[29] Antidepressant therapy for post-traumatic stress disorder (PTSD), apparently mitigating this stress reaction, can increase hippocampal size.[30] The similar effects of music and meditation on our mental health are illuminating.[31]

One reason meditation is seeing a surge in popularity is in direct response to a sense of information overload. Our high-tech world seems to be pushing and pulling us along faster and faster. Holmes offered us some advice for this malaise. He recommended we treat our brain like a well-organized attic space. "I say now, as I said then, that a man should keep his little brain-attic stocked with all the furniture that he is likely to use, and the rest he can put away in the lumber-room of his library, where he can get it if he wants it."[32] He was suggesting we avoid cluttering our mind with unimportant details.[33] As much as he valued knowledge, it had to be useful—everything else he could just look up in his library. Now, to our detriment, our little brain-attics have become congested with excess. Well before the computer age, Thoreau lamented that our lives were being "frittered away by detail."[34] Technological progress doesn't always give us more time; rather, the opposite is often true. In his day it just came in different forms. "We do not ride on the railroad," he wrote, "it rides on us."[35] Well, the internet is riding on us. Boredom has become a rarity. Its empty space is an important time for thoughts to aimlessly roll around in our heads. These down times are when our best ideas come. They surface during a long run, or in the shower, when our minds are not distracted by other things. Lacking boredom, we are now seldom forced to catalyze the imagination through being *stuck*. John Cage wrote about the acquired taste of doing nothing: "If something is boring after two minutes, try it for four. If still boring, then eight. Then 16. Then 32. Eventually one discovers that it is not boring at all."[36]

It is thus unsurprising to see creativity in decline. This contrasts with the trend seen with general intelligence. James Flynn reported that, although IQ is always renormalized to a mean of 100, the raw scores have been increasing over time. People are getting smarter. In his honor, this phenomenon is called the Flynn Effect.[37] Testing for creativity is accomplished with the Torrance Test of Creative Thinking (TTCT). But the raw data here tells a different story. Since 1990, researchers have seen a steady decrease in innovative thinking.[38] This, of course, coincides with the widespread adoption of computers and the internet. Research scientist Janis Whitlock calls the modern world a "cauldron of stimulus."[39] How can we focus on the present when we are perpetually inundated with data, and consider wasting time a sin and multitasking a virtue? Time is money; waste not, want not. It's the Red Queen in Lewis Carroll's *Through the Looking-Glass*— "Faster, faster!"

"The twentieth century is, among other things, the Age of Noise," wrote Aldous Huxley. "Physical noise, mental noise, and noise of desire—we hold history's record for all of them."[40] With all these distractions, it is more important than ever to learn how to be mindful of the present. Holmes accomplished this by scratching away on the violin in snippets of meditation. His pipe, too, was meditative. He was careful, though, to never mistake mindlessness for mindfulness.[41] He recognized that there is a world of difference between passive and active thought. Only one is deliberate. In *A Scandal in Bohemia*, Holmes chided Watson about not *noticing*:

> Holmes: "You see, but you do not observe. The distinction is clear. For example, you have frequently seen the steps which lead up from the hall to this room."
> Watson: "Frequently."
> Holmes: "How often?"
> Watson: "Well, some hundreds of times."
> Holmes: "Then how many are there?"
> Watson: "How many? I don't know."
> Holmes: "Quite so! You have not observed. And yet you have seen. That is just my point.

Now, I know that there are seventeen steps, because I have both seen and observed."[42]

Mindless consumption of music, too, is hearing without listening. It is not deliberate, and there is no engagement. Oliver Sacks bemoaned the ubiquity of passive background music. "We are enveloped by a ceaseless musical bombardment whether we want it or not," he wrote.[43] Huxley described the same phenomenon a bit more curmudgeonly: "That most popular and influential of all inventions, the radio, is nothing but a conduit through which pre-fabricated din can flow into our homes." He complained that it is not just noise to our ears but also "penetrates the mind, filling it with the babel of distractions."[44] This type of interaction with music doesn't encourage thought or creativity. Contrary to some opinions on the matter, it has even been shown to interfere with concentration.[45] Deliberate listening to music has the power to foster creativity and insight. The dual physical and cognitive aspects of playing an instrument are even more productive. This may be a result of its capacity to promote another creativity stimulus: the flow state. Flow is essentially *meditation with intent*.

F#/G♭: RAUSCH

Rausch

CZECH PSYCHOLOGIST MIHALY CSIKSZENTMIHALYI WAS ONE OF THE FIRST TO study the concept of flow. He was certainly the most influential in popularizing the idea. Flow is that great transcendence of time when you are so focused on the present task that "nothing else seems to matter."[1] It is an optimal state of consciousness where we feel and perform our best.[2] We know it as the aesthetic rapture of musicians, or the athlete in the zone. There is complete awareness and pure presence. Nothing external intrudes. Time disappears.

Flow is a profound source of epiphany and transcendence. Author and flow guru Steven Kotler notes that it is stimulated by many of the same qualities that define great scientists: curiosity, passion, and purpose.[3] Flow sits smack in the middle of the continuum between arousal and control. The task must not be so hard as to evoke anxiety, but not so easy that we can perform on autopilot. Flow is inherently active. Passive activities like television and background music do not trigger flow and will not improve cognition or creativity—they probably do the exact opposite. When Holmes contemplated a challenging case, he actively sought flow's epiphanic qualities. Sometimes he employed dubious methods, like cocaine, in his pursuit of flow. In *The Adventure of the Second Stain*, Watson depicted Holmes's strange behaviors as he puzzled over a particularly difficult case:

> All that day and the next and the next Holmes was in a mood which his friends would call taciturn, and others morose. He ran out and ran in, smoked incessantly, played snatches on his violin, sank into reveries, devoured sandwiches at irregular hours, and hardly answered the casual questions which I put to him. It was evident to me that things were not going well with him or his quest.[4]

He was in the thrall of his mind, sifting and sorting, seeking connections, oblivious to the outside world. That is, he was in flow. Leonard Bernstein was grateful when flow aided his composing process. When the ideas come, he said, they come in torrents: "A spark will fly, and I'll be off, sailing, my ego gone. I won't know my name. I won't know what time it is. Then, I'm a composer."[5] Composer Kyle Gann wrote about his own similar experiences in the foreword to John Cage's *Silence*: "It evokes the state of right-brain absorption that can take place during composing, in which the passage of time ceases to be noticed, and logical concepts fall away, in which the will is moving spontaneously and without charts or crutches."[6]

FIGURE 20: LEONARD BERNSTEIN
COMPOSING, SEEKING CONNECTIONS.

Flow is intimately bound to curiosity and amusement. With these preconditions, work feels more like play. This is what professor James Carse meant by "infinite games."[7] There is no end to the possibilities of where it might lead; there is no impossible. William James observed that we notoriously perform below our potential. He called this our "habit of inferiority." Writing in *The Energies of Men*, he said, "As a rule, men habitually use only a small part of the powers which they actually possess and which they might use under appropriate conditions … The human individual thus lives usually far within his limits."[8] How do we rise to the level of our expectations, rather than (as Archilochus condemned) falling to the level of our training? This is where flow comes in. But to get there we need to cultivate the right conditions. According to Csikszentmihalyi, the requisite conditions for flow are as follows: complete concentration, immediate feedback, time expansion or contraction, intrinsic reward, effortlessness, balance between skill and challenge, loss of self-awareness, and a sense of control over the outcome.[9] What do all these qualities have in common? They all suppress frontal lobe function in the brain, a phenomenon called *transient hypofrontality*. The frontal lobes act as a moderating voice of reason, advising the rest of the brain to be cautious. Neuroscientist John Kounios refers to them as our "cognitive blinders."[10] Like those put on a horse to keep it from getting distracted, they help us stay focused and on task. But if it's innovative thinking we're after, we need to take off those blinders. They restrict the breadth our thinking and are an impediment to creativity. A good illustration of this is the fact that patients with damage to their frontal lobes actually perform better on tests of creativity.[11] Children fall into this category as well. We get less creative with age as our maturing brains fall more and more under the "responsible" influence of our frontal lobes. The cost is our creativity.

António Egas Moniz received a Nobel Prize for his invention of the frontal lobotomy. Now, of course, the procedure is considered mutilation, and efforts are being made to de-Nobel Moniz.[12] Fortunately, we have other tools—short of a frontal lobotomy—that can bypass these quiet moderators. Meditation, as discussed in the last chapter, can suppress frontal brain activity in favor of a more creative pattern of thought. Flow may be even more effective at accomplishing this shift. By removing the blinders, it allows the unconscious mind to take control. Recall how much more powerful the unconscious mind is than the conscious one. More parts of the brain can now engage in the creative process.[13]

Csikszentmihalyi may have been the first to formally study flow, but the concept has been understood for a long time. It was particularly influential among the early German thinkers, who employed the term *Rausch* in place of flow. In one sense of the word, *Rausch* implies intoxication and drunkenness. But these philosophers intended its more flow-like meaning: a state of euphoria, a thrill, an exhilaration. Johann Wolfgang von Goethe described the ecstatic joy that followed the "erasure of the separation between high and low."[14] The Grimm brothers described it as the "soul's drunkenness, the delight of the inner faculties that lead to self-forgetfulness."[15] Nietzsche adopted *Rausch* as one of his central precepts. Its power came from its ability to sweep away the boundaries and limitations of consciousness. His *Übermensch* represented exactly this transcendental human, one who had superseded mere existence.

For Nietzsche, *Rausch* was deeply aesthetic, a potential antidote to nihilism and pessimism. Schopenhauer felt it was even more essential, inspiring revelations about the true nature of reality. He envisioned two potential pathways to transcendence: one being physical action and movement, the other music.[16] Playing a musical instrument uniquely accomplishes both. Music embodies movement. Even just listening to music creates an almost involuntary craving to move: toe tapping, head bobbing, and dancing are driven by deeply ingrained connections between the ears and the movement centers of the brain. The visual arts have their virtues, but they cannot match music's engagement of the entire mind and body. As a result, they are not as likely to promote flow.

Violinist Midori Gotō exemplified the concentration required of a virtuosa when, at the age of fourteen, she played Leonard Bernstein's *Serenade for Violin and String Orchestra* at Tanglewood. During the raucous fifth movement (*allegro molto vivace!*), the thin E string on her three-quarter-sized violin snapped. Unfazed, she turned to the concertmaster, who quickly handed over her own full-sized Stradivarius. Without missing a beat, Midori played on. Violinists out there can imagine the challenge of hitting precise notes on this slightly larger instrument. Unbelievably, as she played on, the E string broke again. This time, she was quickly handed the second violinist's instrument. As the *New York Times* reported, "When it was over, audience, orchestra and conductor-composer joined in giving her a cheering, stomping, whistling ovation."[17]

"What could I do?" Midori asked. "My strings broke, and I didn't want to stop the music." It was pure aesthetic rapture: flow, dopamine, the works. Not only did she demonstrate incredible maturity, but she pulled off an amazing feat of concentration and poise. Performing requires all your focus. Midori could not afford to think about whether it made her happy, what she looked like, or what the audience was thinking. She only knew to play with sheer, autotelic joy.

Besides flow, there are other ways the subconscious can make its voice heard. Dreams are one such way. As Heraclitus wrote, "Even a soul submerged in sleep is hard at work and helps make something of the world."[18] Stravinsky believed his dreams were messages from God. The two of them apparently spoke frequently: the *Octuor* came in a dream,[19] as did the *Rite of Spring*[20] and the *Soldier's Tale*.[21] When he struggled with a particular interval while composing *Threni*, he solved the problem in a dream in which he visualized himself in the library looking up the solution.[22] Feynman also recognized the potential of this unexplored frontier. He spent months attempting to record his unraveling stream of consciousness at the precise moment of falling asleep.[23] Dreams are a sorting process: the brain sifting through the day's detritus, making connections, and storing what is important. It helps keep the attic clean. It is one reason why good-quality sleep is so vital to creative work.

German physician Otto Loewi studied the ways neurons in the brain communicate. He studied music as a child and dreamed of getting a degree in art history, but his father talked him into medical school instead.[24] Nonetheless, his first love remained the arts. At the time, it was still unknown whether nerve communication was electrical or chemical. Loewi thought it was chemical but couldn't figure out how to test his hypothesis. Finally, the idea for an experiment came to him in a dream. Awakening, he quickly scribbled notes to himself before falling back asleep. In the morning—*to his horror*—he could not read the notes at all! He fretted and stewed all day, trying to remember, but to no avail. Astonishingly, the muse visited again that night. He wrote his dream down more carefully this time. The next morning, he placed two frog hearts in different chambers of liquid. The first heart was still attached to the vagus nerve. This chamber had a connecting tube into another chamber containing a second heart. Fluid from the first chamber could flow into the second. He then electrically stimulated the vagus nerve, causing the first heart to slow down. After a delay, the second heart also slowed down. This proved that electrical stimulation of the vagus nerve released a chemical that then flowed into the second chamber. He called it *vagusstoff*.[25] We now know it as acetylcholine.

Incidentally, Wallas's four stages of creativity were derived from Poincaré's original five. A particular experience he had as a young scientist made quite an impression on Poincaré and triggered an enduring fascination with the science of creativity. At a conference in Paris, many years later, he shared the story of a sleep-deprived delirium brought on by the muse's caffeine-infused ichor:

> One evening I took a cup of coffee, contrary to my habit; I could not get to sleep, the ideas surged up in a crowd, I felt them bump against one another, until two of them hooked onto one another, as one might say, to form a stable combination.[26]

By morning, he finally understood the Fuchsian functions. He had struggled with them for more than two weeks without sign of progress, but he was just putting in the work necessary to eventually make a breakthrough. All the while, his unconscious mind was sifting, sorting, and looking for patterns that finally emerged from the mist once he was able to take the blinders off his logical mind.

Hard work lays the groundwork for the divine muse. "Chance favors the prepared mind." Mastery, when combined with curiosity, not only facilitates out-of-the-box thinking but builds a bigger box. It's the box, after all, that makes you smart, and these scientists had worked hard to build large boxes. They just needed to let their subconscious play with what they already knew in order to reveal new connections. Intuition, Einstein believed, grew from the well-tilled soil of previous experience.[27]

Once, when the poet Saint-John Perse visited Einstein at Princeton, the two discussed the vagaries of the muse. French writer André Maurois was on hand, and later shared the conversation. When Einstein asked him how poets get inspiration, Perse explained that it came from the subconscious, through intuition. Einstein immediately grasped the connection with scientific creativity. Forecasting Wallas's four stages, he explained how scientific discovery works the same way. It often results from a sudden illumination, he said, that pushes the imagination in a great leap forward. Maurois was struck by the similarity. He had long suspected that a scientist's intuition "stems from an unconscious statistical recollection or from the sudden glimpse of an analogy."[28] The subconscious mind clearly has tremendous power. Music, as a predictable trigger of flow, transient hypofrontality, and meditative insight, is uniquely suited to awaken our extraordinary oration.

D♭:

Atheoretical Cartography

COMPOSER JOHN CAGE SHOOK THE FOUNDATIONS OF OUR PRECONCEIVED notion of music. Cage was a revolutionary, but he worked from inside the system he was to subvert. He studied for a while under Arnold Schönberg. Later, he criticized Schönberg's twelve-tone method for having no zero: "There is not enough of nothing in it."[1] As Susan Sontag wrote in *Styles of Radical Will*, "Silence remains, inescapably, a form of speech."[2] Cage famously pushed even these bounds with his song *4:33*: the length of silence. This is a bit of a misnomer because Cage believed "there is no such thing as silence." His idea was to bring to awareness all the ambient noises we often take for granted, such as a cough, a sneeze, or a shuffling chair. There is never total silence. He frequently mentioned his experience in the Harvard anechoic chamber where, even in total silence, he still heard two sounds: one high pitched and the other low. These, it turned out, were his nervous system and his cardiovascular system—playing Laënnec's melody.[3] Duration was the one musical quality that sound has in common with silence, and Cage loved experimenting with it. In *One10*, a violin's single note is repeatedly sustained for nearly forty seconds at a time, interspersed with long rests of silence. One begins to lose track of which part is the music and which part the silence.

Physician and poet William Carlos Williams wrote in his epic poem *Paterson*, "Dissonance / (if you are interested) / leads to discovery." Cage explored music's boundaries with sound, as well as with silence. He stubbornly insisted that *any* sound can be music. He reinforced this point through the strategic placement of various noises like sirens in his music.[xiii] Cage espoused an ethic of "purposeful purposelessness."[4] He wondered, for instance, "Which is more musical, a truck passing by a factory or a truck passing by a music school?"[5] Cage wanted to shake us out of our ossified habits and force us to listen again. It is the same theme that inspired his work at the boundary between randomness and imposed structure. He went so far as to introduce pure chance into his compositions—sometimes a roll of the dice, other times consulting the *I Ching*, or even following irregularities in the paper he was working on.[6] The result is that we are forced out of autopilot and made to pay attention.

[xiii] Cage's sirens were of the modern, grating variety, distinctly different from those originally envisioned by physicist Hermann Helmholtz. Helmholtz, who was also a composer and pianist, named his invention after the sirens of Greek mythology because he thought its beautiful sound would woo people into paying attention to it! He also, incidentally, built a special piano for experiments in tone production, reminiscent of Cage's prepared pianos. (Root-Bernstein, "Music, Creativity and Scientific Thinking," 64.)

FIGURE 21: COMPOSER AND MUSIC PHILOSOPHER JOHN CAGE.

Notwithstanding Cage's definition, music generally has at least some structure to it. Musicologist Viktor Zuckerkandl argued, "What we hear when a cat runs over the keyboard … is presumably not a melody." Rather, he continued, "A melody is a series of tones that make sense."[7] Only the structural relationship between tones creates music, and these patterns are more of a *gestalt* than a definable entity.[8] This relates directly to Cage's exploration of silence. Pianist Artur Schnabel made this distinction: "The notes I handle no better than many pianists," he wrote, "but the pauses between the notes—ah, that is where the art resides."[9] Like chiaroscuro for the ears, the shadows and empty spaces are as important as the details. This is employed to powerful effect in Beethoven's Fifth Symphony, where the very first note of the theme, an eighth note rest, is "accented silence."[10]

The gestalt structural vagaries of music are one reason why artificial intelligence (AI) has not been very successful at creating any "good" music. No single note constitutes music. Too much variety, and we lose the theme; too little, and we lose interest. Harari sees AI created music in our future, though. At the Yalta European Strategy conference, he hypothesized, "If we think about art as kind of playing on the human emotional keyboard, then I think AI will very soon revolutionize art completely."[11] It is reminiscent of George Orwell's "special kind of kaleidoscope known as a *versificator,*" which mechanically composed catchy music for the masses in his book *Nineteen Eighty-Four.*[12] Perhaps AI will eventually succeed in appealing to our curiosity. It is proving in some areas to be even more creative than humans. Online chess programs, for instance, now monitor for cheating by looking for moves that are *too* highly creative.[13]

But other experts in AI, like Douglas Hofstadter, are not so optimistic. Hofstadter, a pianist and composer in his own right, has established credibility. For a program to write *human* music, he explains, it would have to wander the world, feeling its way through life's trials and tribulations. It would have to know joy, loneliness, resignation, grief, despair, and heartbreak, not to mention jubilation, serenity, and suspense. It would require "a sense of grace, humor, rhythm, a sense of the unexpected—and of course an exquisite awareness of the magic of fresh creation. Therein, and therein only, lie the sources of meaning in music."[14]

Music is an illusion. It is a trick the mind plays as it creates and imposes structure on what is otherwise just a sequence of sounds.[15] Composers masterfully exploit this auditory legerdemain. Maybe this is where Cage and Zuckerkandl can find common ground. First, let's compare the illusion of music to something we are all familiar with. Visual illusions—like Shepard tables [figure 22] and the Ponzo illusion [figure 23]—betray a bug in the brain's operating system. Even when we know the trick, we still cannot stop seeing it. We do not consider that the same thing occurs in the auditory system. Composers prey upon this manipulated expectation. They produce a melody that disappears when the music slows, as in Chopin's Fantasie-Impromptu in C-sharp Minor, or an unexpected chord that startles us, as in Haydn's *Surprise Symphony*.[16] Mozart was particularly facile at assimilating his listeners as an inextricable part of the magic. In *Don Giovanni*, he wrote the sound of an orchestra tuning up into the score. The listener plays an integral part in making sense of Mozart's music. "Their interests and desires," wrote James Gleick, "help form the context without which the music is no more than an abstract sequence of notes ... Mozart's genius, if it existed at all, was not a substance, not even a quality of mind, but a byplay, a give and take within a cultural context."[17] It leads one to wonder whether the same illusions govern our perception of nature's harmony. Does its beauty reflect nature or our own minds?

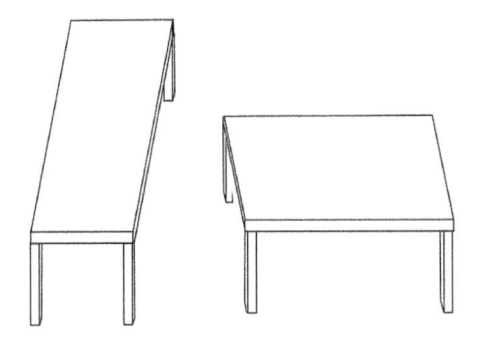

FIGURE 22: SHEPARD TABLES: THE TABLETOPS ARE THE SAME SIZE.

FIGURE 23: PONZO ILLUSION: THE THICK HORIZONTAL LINES ARE THE SAME LENGTH.

Aristotle and John Locke both recognized that our limited senses distort the way we perceive the world. Everything we think we know is colored through the lens of neural recognition and interpretation. Pythagoras was one of the first to appreciate that all sensory perception is musical vibration. He envisioned a vast empyrean symphony played by planetary reverberations. He called it *the harmony of the spheres*. According to Aristotle, "[The Pythagoreans] supposed the whole heaven to be a harmonia and a number."[18] Kepler, in his version of the Pythagorean orchestra, described Earth as an alto. He claimed the Earth sings, "Mi, Fa, Mi: you may infer even from the syllables that in this our home **mi**sery and **fa**mine hold sway."[19] The rest of the choir was filled out by Mercury (soprano), Mars (tenor), and Jupiter (bass).[20]

Undoubtedly, we are oblivious to an entire other world of information. "There is a whole continent of colors redder than red," the physicist Alan Lightman wrote in *The Accidental Universe*, "and another whole continent bluer than blue." Bees see a world of ultraviolet light that we were completely unaware of until the modern technology broadened our horizons.[21] Snakes see the other end of the spectrum in infrared.[22] Sharks navigate with the assistance of electrical fields.[23] It's hard to even fathom what other types of information we might be missing altogether.

We are in a new age where most of our knowledge about nature and the universe now comes from instruments, rather than from our own observations.[24] We have become accustomed to the idea that reality is not what it appears to be. According to physicist Carlo Rovelli, this veil fell when we first realized that the Earth is round and rotates like a "mad spinning-top."[25] What began with Foucault's pendulum has blossomed into the remarkable work at the Laser Interferometer Gravitational-Wave Observatory (LIGO). Sound waves cannot travel through the vacuum of space, but gravitational waves do. Space, if we could see it, ripples like a vast cosmic sea. These waves, bending space-time itself, record the invisible colors and inaudible sounds of events that transpired eons ago.[26] LIGO brings this world to life. We can finally *hear* Pythagoras's great symphony.[xiv]

Beethoven, though completely deaf by age forty-six, could still play with music in his mind's ear. He composed many of his greatest works, including his Ninth Symphony, after the onset of his deafness. His deafness may have been a key factor enabling him to transcend tradition—and the influence of his mentor Franz Joseph Haydn—to create an entirely new period of classical music.

Simply the act of imagining music activates the auditory cortex nearly as much as listening to it. But *playing* music activates it exponentially more. New imaging modalities, like functional MRI and PET, allow us to see how playing a musical instrument coruscates the entire brain in a psychedelic light show. Playing music is unique in its ability to engage the entire brain—you would be hard-pressed to find an area not involved. The visual cortex is involved in reading music. Mirror neurons coordinate play with other musicians. The cerebellum manages rhythm and meter. The motor and sensory cortices are necessary, respectively, to play an instrument with dexterity

---

[xiv] For an example, visit: https://www.youtube.com/watch?v=TWqhUANNFXw.

and to receive tactile feedback from it. It's like a game of "Dem Bones": the auditory cortex connected to the cerebellum; the cerebellum connected to the motor cortex ... Further down the chain, the temporal lobes are brought into action to assess musical pattern. Then the hippocampus discerns whether we have heard this song before. Even the frontal cortex is activated by music, particularly a small collection of cells called the pars orbitalis. This area is the storyteller of the brain. It crafts the story we imagine being told and, by doing so, helps manage our expectations. This is why the surprise note catches our ear, or dissonance irritates.

When experts look at brain scans, most of us look pretty much the same. They cannot tell, for instance, which brain belongs to a doctor, a lawyer, a teacher, a painter, a mathematician, or a poet. But music's effects on the brain are extraordinary; it is easy to identify the brain of a musician. Most notable is an enlarged corpus callosum. This is the sole pathway connecting the two halves of the brain. Its anomalous size is most evident in those who began musical training at a young age.[27] Another area, the planum temporale in the auditory cortex, is also enlarged—especially in musicians with perfect pitch.

Like a conductor for the brain, music coordinates both hemispheres to work synergistically. Functional MRI and PET thus show more symmetrical activation in musicians.[28] Normally, the right hand engages the left brain, and the left hand the right brain. All this traffic runs across that interhemispheric highway, the corpus callosum. Playing music is like rush hour, especially with instruments that employ both hands.

The motor and sensory cortices are organized like a mini caricature, a homunculus [figure 24], embodying all the parts of our anatomy. The part of the cortex controlling the hands is enlarged in musicians in general. There are further, subtle differences depending upon the dexterity and handedness of their instrument.[29] Each of a pianist's hands, for example, performs roughly the same *kinds* of movements. Moreover, the usually dominant right hand typically plays the more complex movement. A violinist's hands, in contrast, have two completely different movement patterns. The dominant hand manages the bow, while the nondominant hand performs the more rapid and precise fingerboard movements. A violinist's homunculus has a huge left hand, so large it can even be seen on MRI.[30] The fact that the violinist's nondominant hand performs the more complex movement may explain its comparatively greater right brain recruitment and bilateral interconnectivity.

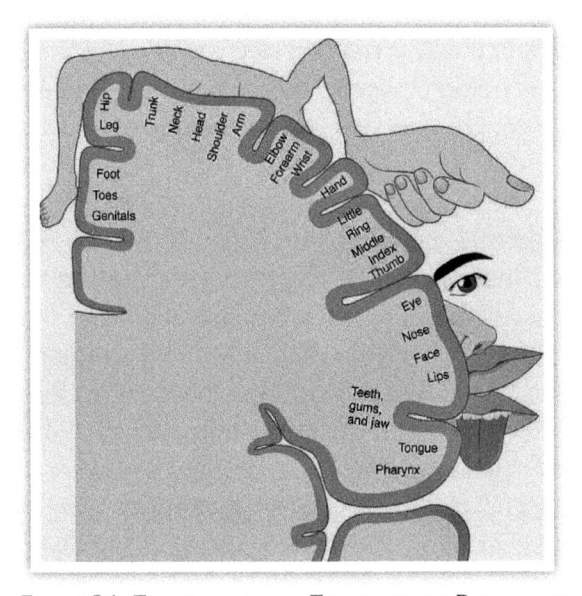

Figure 24: The homunculus. The alchemist Paracelsus first used this term, but in a completely different way than we do today. The alchemists (like the preformationists) believed that each sperm contained a fully formed human that would eventually grow into a full-sized person. Modern scientists have co-opted his term to depict a tiny and distorted human laid out in the brain, sculpted to reflect the corresponding part of the sensory or motor cortex.

It makes sense why so many brilliant scientists and thinkers have been string players. Prior to the violin and piano, the lute and lyre played similar roles. The similar manual patterns would present equivalent neural stimuli. Engaging the right brain influences pattern recognition and problem-solving. Most importantly, it helps us see beyond our usual paradigms. Both Einstein and Holmes were right-handed: Einstein wrote with his right hand, and Holmes, in *The Sign of Four*, kept his revolver in the right-hand pocket of his jacket.[31] For these presumably right-handed, left-brained, and highly analytical minds, it would have been very important to have techniques to engage their right brains in creative problem-solving.

The great Galen, physician to the Roman Emperor Marcus Aurelius, thought the two halves of the brain were simply mirror images.[32] Hemispheric specialization was not suspected until 1780, when the Dutch physician Meinard Simon du Pui called humans *Homo duplex*. He believed the two hemispheres functioned independently, like two separate brains: "a right man and a left one."[33] Descartes suspected the pineal gland connected the two hemispheres. It seemed appropriate to him that the seat of the soul should unite these two different minds. Dr. Roger Sperry finally recognized that each hemisphere performed specialized tasks, and that they were connected by the corpus callosum. Treating patients with intractable epilepsy, he severed the communication pathway between the two hemispheres. Surgical commissurotomy severs all communication between the hemispheres, helping stop the spread of seizure activity, and had been shown to be effective as far back as far back as 1940.[34] Sperry described different behaviors, depending upon whether the patient was thinking with his or her right or left brain. "The same individual can be observed to employ consistently one or the other of two distinct forms of mental approach and strategy, much like two different people, depending on whether the left or right hemisphere is in use."[35] This discovery won him the Nobel Prize in Physiology or Medicine in 1981.

The left side, he found, tends toward the abstract and verbal, while the right side is more perceptual. The left deals with knowns, and the right with unknowns. Without the right brain, we would act like children's book character Amelia Bedelia, our left brain interpreting everything so literally we could hardly function. We would struggle with figurative speech, jokes, gist, and music. But without the left side, we would struggle with language, focus, and analytical thought. The actions of each hemisphere were so discrete that Sperry concluded that each possessed its own consciousness.[36] This dual consciousness hearkens back to du Pui's *Homo duplex*, but in Sperry's case it was appropriate, because his patients literally did have two separate brains.

Left and right patterns are not just biological constructs. They are abundantly present throughout the universe. About half of known galaxies rotate to the right, or clockwise, while the other half rotate to the left. Whereas Venus and Uranus have right-handed orbits, the other planets are left-handed. These patterns percolate all the way down to the microscopic level. Every organic molecule has both right-handed and left-handed versions. Curiously, every amino acid on Earth is left-handed, and every sugar is right-handed.[37] Although chemists can easily make mirror-image molecules, they cannot be used by living systems. Our bodies will only metabolize molecules with the proper handedness.

Arguments such as these, contradicting a natural and godly symmetry, were initially hard for physicists to swallow. Einstein's friend and colleague Wolfgang Pauli was willing to bet that the Lord (that is, nature) was *not* a left-hander. Chen Ning Yang and Tsung Dao Lee proved him wrong and won a Nobel Prize for their efforts.[38] Nature is not symmetric, and she does distinguish left from right, even at the subatomic level.

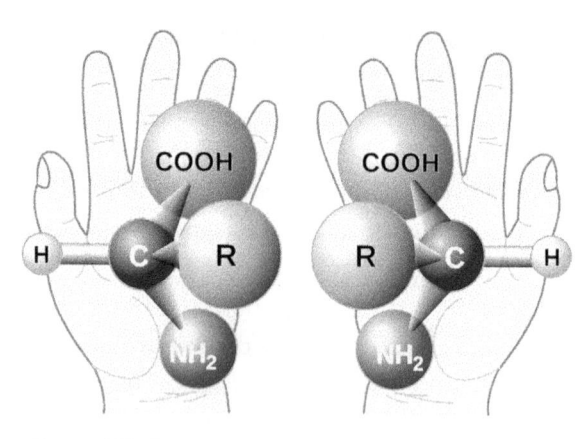

FIGURE 25: A CARBON MOLECULE CAN FORM BONDS WITH FOUR OTHER ATOMS. IF THE ATOMS IT ATTACHES TO ARE ALL DIFFERENT, THEN THE RESULTING MOLECULES CAN BE MIRROR IMAGES OF EACH OTHER. RIGHT- AND LEFT-HANDED MOLECULES HAVE THE SAME CHEMICAL MAKEUP BUT CANNOT BE SUPERIMPOSED BECAUSE OF THEIR MIRROR IMAGE CHIRALITY.

The dualism of Descartes may have fallen, but the complementary functions of the right and left hemispheres are vital to a vibrant interaction with the world. As we map all the various brain regions, what psychologist and neuroscientist Daniel Levitin calls "so much atheoretical cartography,"[39] we risk losing sight of the vast and complex interconnectivity. This is where the mind's real power lies. Pattern, it turns out, is not an anatomic entity. It exists solely as a relationship between entities, neither tangible nor mappable.[40]

From a teleological perspective, this convoluted arrangement enhances its evolutionary value. Take the auditory system for example. It evolved out of the vestibular system, our balance and movement center. Hearing and balance are thus colocated in the inner ear. Experiments that incorporate dichotic listening—placing different auditory inputs in each ear—highlight the divergent roles of each cerebral hemisphere. For example, a song heard through the left ear is remembered better than one heard through the right ear. The opposite is true of the spoken word. Orientation in space is also comanaged by the vestibular and auditory systems. Just as visual depth perception depends on two eyes, auditory directional perception depends on two ears. We can usually localize the direction a sound came from to within about three degrees. Bats and owls, the champions of stereophonic hearing, can pinpoint it within just one degree.[41]

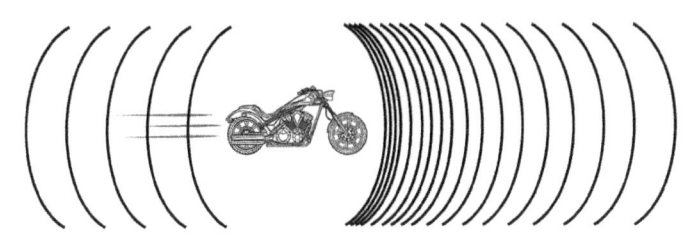

FIGURE 26: THE DOPPLER EFFECT. AS THE MOTORCYCLE DRIVES TO THE RIGHT, IT CATCHES UP WITH ITS OWN SOUND WAVES, RESULTING IN A HIGHER FREQUENCY.

The remarkable phenomenon of stereophonic hearing is made possible by the Doppler effect. The classical example of this occurs when a motorcycle passes you on the road. An approaching motorcycle sounds different than one driving away from you. When it drives toward you, driving *into* the sound waves approaching your ear, the waves are compressed. The moment it passes you, the sound changes. Now the motorcycle is driving away from its sound waves, stretching them out, at least in reference to your ear [figure 26]. To the motorcycle rider, the waves remain the same throughout this sequence, and so she never hears the change—another auditory illusion. Your brain, which can only interpret your own ears' inputs uses the wavelength discrepancy to identify directionality. This change can literally alter the notes heard.

Doppler's conjecture of 1842 was confirmed by experiment a few years later. Christophe Ballot, a Dutch meteorologist, lived next to a train line and had already recognized the change in pitch of the whistle as the train zoomed past. For his experiment, he filled an open train car with trumpet players and had a control group on the platform. As the train approached the platform, the musicians on the train sounded sharp. Then, as they passed by, the brass band suddenly sounded flat.

The same thing happens with light. Remarkably, Doppler's original report alluded to color changes in stars rather than sound waves. This phenomenon explains the famous red shift, first noticed by Edwin Hubble. Light streaming away from the viewer is spread out into longer wave lengths: towards red. A blue shift would indicate light moving toward us. Hubble's discovery led to the uncomfortable realization that everything in the universe, in every direction, is expanding *away* from us—crucial supporting evidence for the big bang theory. Einstein took this thought experiment even further. Relativity suggests that the same thing happens to time itself.[xv]

Quantum mechanics taught us that the whole universe is vibration. Indeed, all perception is vibration. Our ears are attuned to wavelengths between 20 and 20,000 hertz (Hz). Lower frequencies are processed predominantly by the right brain, while higher pitches—because they are more closely related to speech and language—disproportionately activate the left side.[42] A piano's eighty-eight keys span seven octaves plus a minor third and range from 27.5 to 6,000 Hz, nicely reflecting our auditory range.

Isaac Newton was not particularly musical. He famously walked out of the only opera he ever attended. But he had enough background from his childhood music studies to recognize the similarity between the musical scale and the colors of the spectrum.[43] He commented on how precisely the rainbow's seven colors correlated to the seven notes of the octave. The parallel evolutionary origins of our sense organs

---

[xv] Though admittedly imprecise from a technical standpoint, it works as a cognitive analogy to simplify the concept. (Brown, *Planck*, 61-62)

are highlighted by the fascinating relationship between color and sound. Tune up the lowest key on the piano by forty-three octaves, and you eventually get red—at 430 trillion Hz. Go further, all the way to 750 trillion Hz, and it becomes violet. Light, like sound, is technically colorless. Tone, whether color or sound, only exists in the brain's analysis of vibration and oscillation. A falling tree makes no sound if there is no brain to interpret its waves of atmospheric disruption.

A♭:

Biomimicry

THIS IS A GOOD TIME TO REMIND OURSELVES THAT HUMANS ARE NOT THE ONLY animal with musical abilities. Attempting to trace our innate musical talent back to some prehuman ancestor is an impossible ask. There is just no fossil record of something so intangible as music. It is hard enough to sort out its evolutionary influence in ancestral hominids. We can only make educated assumptions based on what we know today. Though there are similarities between human and animal music, that alone does not imply a common ancestry. It could just as likely—probably more than likely—indicate convergent evolution: a trait that has developed more than once, in different classes of animals, because of its selective advantages. Flight wings, for example, developed independently in bats, birds, and insects.

Richard Wagner's household held a veritable menagerie. His dog, a spaniel named Peps, used to lie under the piano while Wagner composed. The dog's howl would alert him if Peps did not like the composition. He also had a parrot. The story goes that Wagner's wife, Minna, once taught it to say, "Richard Wagner, *du bist ein grosser mann!*" She was determined that, if not her, at least one member of the household would appear to appreciate her husband's genius.[1]

FIGURE 27: FIDDLEHEAD FERNS—THEIR SHAPE ALSO FOLLOWS THE FIBONACCI SPIRAL'S PATTERN.

Erik Satie observed, "We cannot doubt that animals both love and practice music. That is evident. But it seems their musical system differs from ours. It is another school."[2] At the risk of anthropomorphizing, *Homo sapiens* is probably not the only animal that *appreciates* music. Aside from that of humans, the most complex music in the animal kingdom belongs to the baleen whale. Evolution has endowed them with several operatic traits. The laryngeal organ, for example, allows them to sing without exhaling—a necessity if you want to do so underwater. The range of their voice almost perfectly mirrors that of the piano, spanning just over seven octaves.[3] Novel compositions employ rhymes, themes, and repetition in a repertoire that evolves over time.[4] Most songs last about thirty minutes, although one whale was heard to sing for twelve hours straight—maybe he had an earworm!

But here's the rub: Do whales perceive their music as

beautiful? Do they *appreciate* it? And is it really music if it is not consciously enjoyed? I think these are straw-man arguments. They are untestable pseudoscience. Are we not anthropomorphizing if we think animals need to appreciate music in the same ways we do? Early human music was chiefly communicative, and likely played a role in courtship—the same roles we attribute to animal music today. Whales have different sounds for echolocation, feeding, and courtship. It is tempting to acknowledge that they may at times also sing for sheer joy. Why should humans be the sole beneficiaries of this remarkable tool?

In *The Adventure of the Solitary Cyclist,* Watson narrated, "Holmes and I walked along the broad, sandy road inhaling the fresh morning air, and rejoicing in the music of the birds and the fresh breath of the spring."[5] John Cage was also thrilled by birdsong. Once, on a walk, he was excited by a merle of blackbirds taking flight with "a sound delicious beyond compare."[6] Birdsong seems as if it were specially created especially for human ears—most of it falls within our peak hearing range, between 2,500 and 3,500 Hz.

Humanity's esteem for birdsong is unrivaled, and it often even inspires our own music. The Koyukon live in northern Alaska and the Yukon, an ancient Athapaskan people with language ties to the southwestern Navajo and Apache. Their music mimics the song of the loon. A Costa Rican tribe copies the notes of a riverside wren with a flute.[7] What Lucretius fondly called "*liquidas avium voces*"[xvi] has been shamelessly exploited by Western composers. The opening of the rondo in Beethoven's Violin Concerto in D Major mimics the European blackbird. The second movement of his *Pastorale* features the flute as nightingale and the oboe as quail, while the clarinet toots the "unmistakable falling third of the cuckoo."[8] Vivaldi's flute concerto is appropriately called *Il Gardellino* (the goldfinch).

[xvi] "Flowing bird voices," from *De Rerum Natura.*

Mahler's first symphony recalls the cuckoo, Respighi's *Fountains of Rome* again employs the nightingale, and Schubert's *Quail Song* features the you-know-what. Finally, the synesthete composer Olivier Messiaen mimicked the entire dawn chorus in *Chronochromie*. Richard Powers referred to Messiaen's *Quartet for the End of Time* as "birdsong's answer to the war,"[xvii] with the clarinet as blackbird and the violin as nightingale.[9] We are repeatedly, it seems, attracted to the nightingale, poor Philomela, the lover of music.

FIGURE 28: THE NIGHTINGALE, GENUS PHILOMELA ("LOVER OF MUSIC"), IS NAMED AFTER THE TRAGIC GREEK HEROINE OF THE SAME NAME. HER BROTHER-IN-LAW, KING TEREUS, RAPED HER AND CUT OUT HER TONGUE TO PREVENT HER FROM TELLING HER SISTER. THE GODS LATER TURNED HER INTO A BIRD TO AID HER ESCAPE FROM THE ANGRY TEREUS AFTER SHE GOT HER REVENGE (IN A GRUESOME MANNER, I MIGHT ADD). THE FEMALE NIGHTINGALE DOESN'T SING—ONLY THE MALES STILL HAVE THEIR "TONGUES."

[xvii] The *Quartet for the End of Time* was composed while Messiaen was a prisoner of war at Stalag VIIIA during World War II. The story of its composition and eventual performance is worth reading. For an in depth read, see Rischin, Rebecca. *For the End of Time: The Story of the Messiaen Quartet*. Ithaca, NY: Cornell University Press, 2003.

Aside from a few insects, birds are the only animal that knows how to count. They are fully aware of how many eggs they laid and whether one has gone missing. Perhaps this is another example of the overlapping skills of arithmetic and music. Birdsong shows all the sundry mathematical variations of human music: pitch, tempo, theme variations, key changes, even transposition and harmony. Various species employ different scales: the canyon wren sings in the chromatic, the wood thrush in the diatonic, and the hermit thrush in the pentatonic.[10] The Socorro mockingbird counter-sings like a melodic canon, and musician wrens sing in ternary form, vaguely reminiscent of the sonata, with a theme, variations, and finally a restating of the original theme.[11]

Birdsong is at least somewhat genetically determined. Birds isolated for many generations will eventually return to their species' wild-type songs.[12] Some birds sing hundreds of songs, others only a few. Male canaries grow more brain cells in the spring to facilitate more complex songs, which have been shown to be more successful at attracting a good mate.[13] They shrink back to normal size after nesting season is over. But there is also clearly a cultural, or learned, aspect to birdsong. Steller's jays have learned to imitate the call of a red-tailed hawk.[14] As other birds flee the area in fear, the jays gain unfettered access to their food. Lawrence's thrush are perhaps the champion of all imitators. Ornithologists have heard them cover the songs of 173 different bird species, not to mention various frogs and insects. They even integrate other species' songs into their own unique compositions, wooing the ladies with their incredible oeuvre.

Common starlings learn songs by vocal tradition and are also superb imitators. Native to Europe, they were introduced to the United States in Central Park, on March 6, 1890, by a New York pharmaceutical manufacturer named Eugene Schieffelin.[15] His group, the American Acclimatization Society, dreamed of bringing all the birds in Shakespeare's plays to the United States. His previous attempts at introducing bullfinches, nightingales, and skylarks had failed to establish breeding populations, but with the starling he achieved a success beyond his wildest dreams. Little did he appreciate the ecological menace he would inflict. Those original sixty birds have since ballooned into unfathomable murmurations. The North American starling population now numbers well over two hundred million.[16]

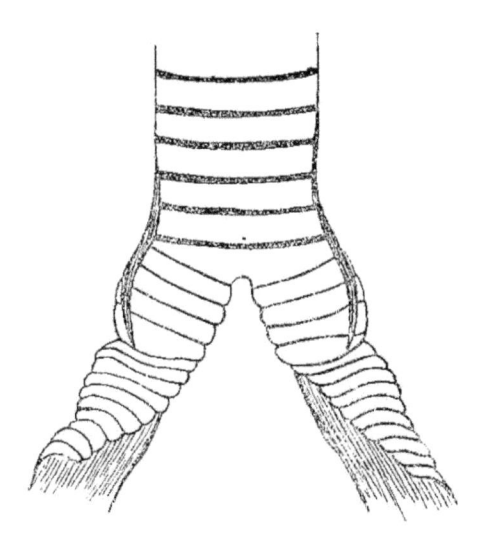

FIGURE 29: CLOSE-UP OF THE AVIAN SYRINX. NOTICE HOW EACH VOCAL ORGAN RECEIVES ITS AIRFLOW FROM A DIFFERENT BRONCHUS.

The syrinx is the avian equivalent of our larynx. Our vocal cords, situated in the trachea, produce sound from only a single stream of air. Birds, uniquely, have a vocal cord equivalent in each bronchial tube. This endows them with a remarkable ability to sing two different parts simultaneously.[17] The starling's skillful mimicry and musical talent enticed Mozart, as it had

Pliny and Shakespeare many years earlier.[xviii] On May 27, 1784, Mozart purchased a starling for 34 kreuzers. It won him over when it whistled part of his recently completed Piano Concerto No. 17 in G. The bird improvised a fermata on the last beat of the first full measure and sang a G-sharp instead of G natural in the following one.[18] Mozart loved its take on his composition: *"Das war schön!"* he wrote in his expense account for the day. "That was beautiful!" When the starling died a couple years later, Mozart wrote a light divertimento he titled *Musical Joke*. Written just eight days after the bird passed, it is full of half-step *mistakes*. The final cadence, written in two separate keys, was almost certainly an homage to his starling.[19]

The very fact that so many animals "practice" music, as Satie said, should give us pause when we consider our own music. It has evolved, biologically and culturally, over the course of human history. But music is also part of nature, innate to many different types of animals. It begs the question as to whether music reflects nature's deeply hidden patterns, as Einstein felt, or merely a pattern imposed by our own minds, a la Kant. Maybe it's both. "Science cannot solve the ultimate mystery of nature," wrote Max Planck. "And that is because, in the last analysis, we ourselves are part of nature and therefore part of the mystery that we are trying to solve. Music and art are, to an extent, also attempts to solve or at least express the mystery. But to my mind, the more we progress with either the more we are brought into harmony with all nature itself."[20]

[xviii] In *Henry IV*, Hotspur considers teaching a starling to say the name Mortimer to disturb the king's sleep.

E♭:
Semiotic Metaphor

MUSIC AND MATHEMATICS ARE BOTH EXAMPLES OF SEMIOTIC METAPHOR: THAT is, they are systems of symbols that seek to represent the real world. Like the shadowy walls in Plato's cave, they attempt to link patterns and abstractions through real, though intangible, methods (tones and numbers, respectively). They aspire to represent macrocosm within microcosm, complexity within simplicity. The symbol, then, transcends reality. It strips away physicality and opens the door to abstraction.[1] "The measure in which science falls short of art," wrote mathematician J. W. N. Sullivan, "is the measure in which it is incomplete as science."[2] The two are so inextricably linked that failure in one realm predicts failure in the other. Why are science and art, math and music, so intimately bound?

Where music possesses mathematical properties, mathematics is itself endowed with aesthetic and musical qualities. Renowned mathematician James Joseph Sylvester noted that music is the "mathematics of sense," and mathematics the "music of reason."[3] The former is "aesthetically inevitable," and the latter "logically inevitable."[4] Both mathematicians and musicians have borne witness to this overlap. Ludwig

Wittgenstein, in high praise, compared Bertrand Russell's *Principia Mathematica* to music.[5] Chopin also saw the two fields as more similar than distinct. His friend, painter Eugène Delacroix, explained:

> The fact is that true science is not that which is ordinarily understood by this word, that is to say, a phase of knowledge different from art. No! Science thus viewed, demonstrated by a man like Chopin, is art itself. And on the other hand, art is then no longer that which the masses believe it to be, that is, a sort of inspiration which comes from I don't know where, which proceeds haphazardly, and which presents only the picturesque exterior of things: it is reason itself, adorned by genius, but following a course that is necessary and controlled by superior law.[6]

Are the patterns we see in music and math just human inventions, or are they displays of an underlying universal harmony? Kant argued for the former, suspecting humankind imposed our own ideas of beauty upon nature. The patterns we see are just so much apophenia, like the conspiracy theorist who finds associations where none really exist. The great physicist Heinrich Hertz, on the other hand, believed otherwise. "One cannot escape the feeling," he wrote, "that these mathematical formulae have an independent existence and an intelligence all their own, that they are wiser than we are, wiser even than their discoverers."[7] Also staunchly in Hertz's camp was English writer Eden Phillpotts. He described our growing wisdom as a magnifying glass focused upon nature's beauty. "The universe is full of magical things," he said, "patiently waiting for our wits to grow sharper."[8]

Mathematicians have long considered their theories *discoveries* rather than *inventions*. Similarly, Stravinsky believed his compositions were messages from God. In this epiphanic state he sensed the close kinship of music and math, enjoying that they both swam "seductively just below the surface."[9] Bearing witness to the powerful relationship between the two fields, he said, "The musician should find in mathematics a study as useful to him as the learning of another language is to a poet."[10] Edward Rothstein studied both mathematics and music and wrote a beautiful piece for the *New York Times* describing his moment of insight into their shared harmony:

> When I worked at learning a Beethoven sonata while also trying to understand the Gödel Incompleteness Theorem, the affinity between these activities was evident. Both mathematics and music create languages that have compelling force in shaping understanding and feeling.[11]

Einstein's friend and colleague, Paul Ehrenfest, experienced a similar sense of congruence between the two fields. As he wrote to Einstein, "When I heard Busch playing for 5 minutes—Beethoven quart. op. 59, no.1, I thought—he must understand mathematics well."[12]

Pythagoras believed that "only through number and form can mankind grasp the underlying meaning of the universe."[13] Hearkening back to van 't Hoff's depiction of "pathological imagination," many of Pythagoras's beliefs bordered on the occult. He considered the amicable numbers and perfect numbers particularly special.[xix] The perfect number six, for instance, symbolized the six days of creation. The perfect number twenty-eight corresponded to the lunar cycle. These deep mathematical patterns in nature suggested much more

---

[xix] *Amicable numbers*, like 220 and 284, have divisors that add up to each other (divisors of 220 add up to 284, and the divisors of 284 add up to 220). *Perfect numbers*, like 6 and 28, have divisors that add up to themselves (28 = 1, 2, 4, 7, 14).

to Pythagoras than mere coincidence—it suggested divine intent. Pythagoras was the first to take notice of the similar mathematical relationships present in music. Both, he preached, facilitated human insight into the hidden structures of nature. Pythagoras was a skilled lyrist, and his curiosity was piqued by the notes he played. He found that an octave was produced by halving the string length. Others must have noticed this before, but none had expressed it mathematically. Before long, his student Philolaus identified that an octave was not divided into two equal parts but rather a fourth and a fifth.

Plato took up where the Pythagoreans left off, but with more scientific rigor. Pythagoreans had stubbornly insisted on a "perfect concordance between things arithmetical and things geometrical."[14] One of Plato's greatest accomplishments was proving this idea fallacious. In doing so, he emancipated geometry from arithmetic.[15] He was also the first to recognize irrational numbers (numbers that cannot be expressed as a ratio) as true numbers. As Philolaus had noted, an octave is split into a fourth and a fifth. There is no halfway point. This led to certain philosophical hurdles for the Pythagorean system. For one thing, it brought to light the problems with irrational numbers, which violated the tenets of divine beauty in their system. When the Pythagoreans attempted to calculate the geometric mean of the octave, they couldn't avoid getting tangled up with the square root of two. Avoiding the cognitive dissonance raised by this little tidbit, they simply suppressed it. Legend has it that the Pythagorean who divulged this secret died in a shipwreck under suspicious circumstances.[16] Plato sought to set the record straight. The mathematician and musician Theodorus, ably portrayed as Theaetetus by Plato, was purportedly the first to prove the irrationality of the square root of two.[17] This sounded the death knell for Pythagorean philosophy.

Theodorus's proof, if it ever existed, was never found. It was left to Euclid, the father of geometry and a student of

Plato's school of thought, to formally prove the irrationality of the square root of two. Euclid also wrote a discourse on the relationship of consonance to whole numbers, further solidifying the ties between music and mathematics. Aside from the Bible, his *Elements* is considered the most reproduced and influential book in the Western world.[18]

Another Greek mathematician, Ptolemy, also highlighted the interchangeability of music and mathematical equations in his book *Harmonics*. Whereas Pythagoras had based his music theory around the ratio of the fifth, 3:2, Ptolemy expanded it well beyond these limitations. Johannes Kepler built on Ptolemy's work in his own *Harmonices Mundi*. Kepler was intrigued by Pythagoras's *musica universalis*, but he abandoned Pythagorean tuning in favor of more geometric ratios. He vitiated two thousand years of tradition when he wrote that, "In fact, there are no real sounds in the heavens, and the motion is not so turbulent that a whistling is produced by friction with the heavenly air."[19] His own concept of harmony encompassed a congruence with nature, as measured by the daily motions of the planets at their nearest and farthest points of orbit around the sun (the perihelion and aphelion, respectively). The duet of Earth and Mars, for example, played a minor tenth at perihelion and a perfect fifth at aphelion. Saturn and Jupiter harmonized with a twelfth at aphelion and an octave at perihelion. Indeed, every pair of planets, save one, had consonant ratios corresponding to musical intervals. The only interval missing was the perfect fourth. Fortunately for his system (God would have never left out the perfect fourth!), Kepler found this interval in the relationship of the moon's orbit as seen from Earth.[xx] Meanwhile, Jupiter and Mars, alone among the planets, were dissonant. They played a disagreeable double octave plus a minor third.[20] Kepler remained puzzled by

---

[xx] Comparing the moon's motion at apogee and its motion at perigee, as seen from Earth, results in a perfect fourth.

this mystery until his death. The cause of this incongruence was ultimately found in the asteroid belt between these two planets. The inclusion of these scattered elements of an unformed planet harmonizes his planetary tuning.[21]

Kepler's contemporary, Galileo, carried out advanced scientific investigations into the properties of vibrating strings. Distinguishing himself from Kepler's more divine sense of harmony, Galileo suspected a numerical explanation. "I have never been able to understand why some combinations of tones are more pleasing than others," he wrote.[22] Eventually, he elucidated the relationship between string length, vibration, and tone. In nearby Cremona, Italy, Andrea Amati was busy creating the world's first violins, but they were not yet widely available. Galileo, naturally, played the lute. His admiration of the lute's "affections of harmony, hardness and softness, harshness and sweetness" evokes later depictions of the violin.[23]

The Swiss genius Leonhard Euler was a titan. "One of the greatest intellectual supermen in the history of mankind,"[24] Euler published more than eight hundred works during his career, amounting to almost a third of the entire science and mathematics literature of the eighteenth century! At parties he was called on to recite Virgil's *Aeneid* by memory. Later in life, blinded by cataracts, he optimistically reassured his friends, "In this way I will have fewer distractions."[25] Euler spent much of his creative energy on music. He developed a grand unification of music and math, indexing chords and intervals by their "degree of agreeableness."[26] At the time, these efforts went largely unappreciated. "It is said," explained one writer, "that his musical theory did not triumph because it was too advanced in mathematical computations for musicians, and too musical for mathematicians."[27]

The behavior of violin strings continued to torment mathematicians until 1746, when Jean le Rond d'Alembert's wave theory finally enabled mathematicians to explain the phenomenon. Rejuvenating the precepts of the quadrivium, it allowed mathematicians to better understand number over space and time. d'Alembert conceptualized mathematical motion as a succession of staccato beats. If the notes are played fast enough, they eventually blur into legato.[28] Wave theory eventually morphed into partial differential equations. The most famous equations in physics, including those of Einstein, Boltzmann, Erwin Schrödinger, and Maxwell, were galvanized by this music-inspired theory.

Numerous mathematical models have attempted to make these shared patterns explicit. Rothstein noted that "Sound is simply heard number; number is latent sound."[29] Mathematicians have devised a motley crew of modalities to study musical patterns and translate them into mathematical formulae: set theory, group theory, and modular arithmetic all attempt to subtly decode the structures of music. A classical pianist may play thousands of notes—*all memorized*—during a performance. It is only through some sort of innate group theory—what other fields refer to as chunking—that they can possibly remember it all.[30]

Different mathematical systems through the ages exhibit varying ways of looking at number in relation to pattern. The Aztecs used a vigesimal counting system based on the number twenty (like the old English score). The Aborigines use a binary counting system with a base of two, and the ancient Babylonians' system was based on the number sixty. Nowadays, the Western world almost universally uses the base ten decimal system. Other counting systems still have their places, though. Modular arithmetic (like mod seven and mod twelve) specifically addresses the musical intervals. Instead of a repeating pattern every ten count, these number systems repeat every seven or twelve, reflecting the seven notes of the major and minor keys or all twelve notes of the chromatic scale. We already, perhaps unknowingly, use mod twelve regularly: our clocks are based on this system. In this system, nine plus five equals two (or, on the clock, 9:00 plus five hours is 2:00). Analyzing a musical score with modular arithmetic brings to light patterns and intervals within its structure.

In Bach's *Art of the Fugue*, it is much easier to see the various transpositions, inversions, diminutions, and augmentations when examined with modular arithmetic.[31] When these patterns get converted into fractals, they become even more awe-inspiring. Benoit Mandelbrot's seminal work, *The Fractal Geometry of Nature*, introduced fractals to the world in 1982. He described fractals as "a rough or fragmented geometric shape that can be split into parts, each of which is (at least approximately) a reduced-size copy of the whole."[32] In other words, no matter how much you magnify or reduce a fractal it will continue to look the same. Its perimeter is infinite.

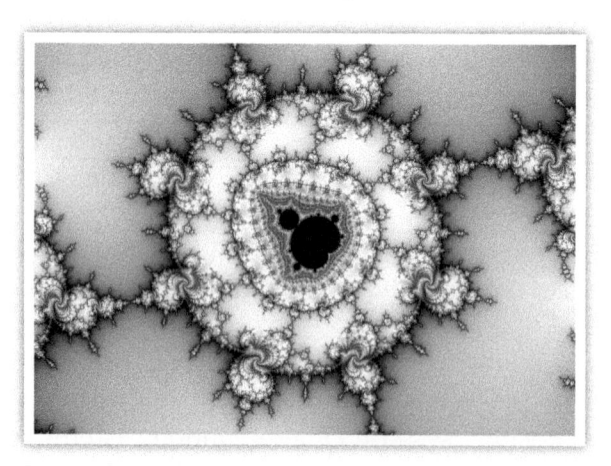

Figure 30: The fractal geometry of Benoit Mandelbrot. Zooming in or out reveals a pattern that is infinitely recurring whether you go deeper or more expansive.

When musicologist Harlan Brothers analyzed Bach's Cello Suite No. 3 with fractals, he found an uncanny similarity. It's almost spooky. As Brothers explained, "The fact that Bach was born almost three centuries before the formal concept of fractals came into existence may well indicate that an intuitive affinity for fractal structure is, at least for some composers, an inherent motivational element in the compositional process."[33] "Fugues from fractals," as Powers writes in *Orfeo*, is not a stretch at all. A "prelude extracted from the digits of pi" may not be far behind.[34] There are inherent mathematical structures in all music. But like Einstein's vain searching for a unified field theory, there is also no mathematical concept that truly *explains* music.

Composer and mathematician Iannis Xenakis has taken the music-as-math relationship even further. Utilizing complex stochastic processes, he has generated musical compositions that are considered excellent. Stochastic processes are essentially based on probabilities, so, in a sense, this technique could be a distant relative to Cage's *I Ching*-inspired compositions. But even here, one must ask whether the computer is truly *composing*. As Hofstadter noted, a computer does not *think* about what it is doing. Attributing the composition of this type of music to the computer, he writes, "would be like attributing the authorship of this book to the ... phototypesetting machine with which it was set."[35]

String theory once held high hopes for resolving Einstein's long-sought grand unification theory. It replaces particles with vibrating strands of energy, the smallest constituents of matter. These one-dimensional strands vibrate like strings on a violin. Different tuning, or modes of vibration, explain the subatomic particles and the four fundamental forces of the universe: the strong (holding atomic nuclei together), the weak (radioactive decay), the electromagnetic, and the gravitational.[36] One curious consequence of string theory is the prediction of an essentially infinite number of alternative universes, all with different physical laws. According to the multiverse concept, the properties of our universe may be (in violation of Einstein's famous statement that "God does not play dice"[37]) merely a result of random throws of the cosmic die.

Henri Poincaré's invention of topology transformed the study of musical patterns and structures. This entirely novel realm of mathematics maps the equivalences among different shapes. It seeks only what is most essential: Two shapes are considered the same if one can be stretched into the other. For example, a circle is like a square, but neither resembles a line. "Mathematicians study not objects," Poincaré wrote, "but relations between objects."[38] How does this relate to music? Chord nomenclature, Rothstein explained, demonstrates that music also does not deal in absolutes. A major triad or a diminished seventh only refers to how the notes of the chord correspond to their roots. "Music is not about things," he wrote, "but about the relations between things."[39] Just like topology.

One unique topological shape is the mobius strip. This ribbon of paper is twisted once on itself before forming a loop. Its surface can be traced perpetually from one side to the other and back. Another example is the torus, or donut. These are distinctly different from other shapes. The sphere of a balloon, for instance, cannot suddenly become a torus as it inflates or deflates. Eighty years before Poincaré, the remarkable chord progression in measures 143–176 of the second movement of Beethoven's Ninth Symphony was written with a chord progression that mathematically forms a torus. Even earlier, Bach's *Musical Offering* had a passage written as if it were on a Mobius strip.[40] These composers, again, wrote with structures that foretold not-yet-discovered mathematical patterns. It is enough to make a skeptic believe that they are a sublime reflection of nature's framework.

The golden ratio has long been revered by artists with an eye for symmetry. Think of the golden ratio as formed by two parts whose ratio to each other is the same as their combined ratio to the larger of the two numbers.[xxi] It was employed throughout the ancient world: Pythagoras considered it the most beautiful of proportions, the Parthenon in Athens was built to these specifications, and da Vinci purportedly used it in his paintings. da Vinci held it in such high esteem that he illustrated Luca Pacioli's book on the subject in 1509.[41] It reflects many of the patterns seen in nature, like the spiral curves of a nautilus shell, the layout of seeds on a sunflower's disk, and even the branching of tree trunks.

The golden ratio can be mathematically explained by a sequence of numbers that has become known as the Fibonacci sequence. In actuality, the sequence was first recorded by Indian mathematician and poet Acharya Pingala over a thousand years before Fibonacci.[42] Generations of Sanskrit poets used the numbers to determine word length and help distribute syllables in the most pleasing way. Fibonacci[xxii]—Italian mathematician Leonardo Pisano—gained everlasting fame when he first recorded the sequence for Western mathematicians. It works as follows: the sum of the two preceding numbers yields the next number in the sequence (0, 1, 1, 2, 3, 5, 8, 13, 21, 34, 55, *etc.*). The ratio of each number to the previous one converges toward a ratio of 1.61803..., the golden ratio. Like pi and other irrational numbers, the decimal places are infinite.

[xxi] The golden mean exists when a line is divided into two parts and the longer part (a) divided by the smaller part (b) is equal to the sum of (a) + (b) divided by (a). $\varphi = \frac{a+b}{a} = \frac{a}{b} \cong 1.61803\,...$

[xxii] Fibonacci derives from the Italian *filius Bonacci*, the son of Bonaccio.

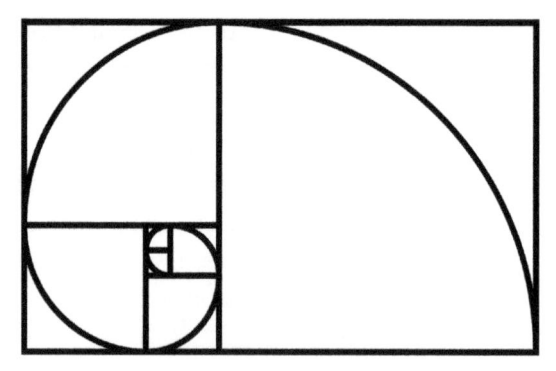

FIGURE 31: THE ARRANGEMENT OF SEEDS PERFECTLY FOLLOWS THE FIBONACCI SPIRAL.

FIGURE 32: SPIRALS IN A SUNFLOWER HEAD.

Musicians also value the golden ratio. Stradivarius made his violins to these specifications. A piano has five black keys, eight white keys, and thirteen total keys in an octave. Debussy structured many of his middle period works in this way, and—like earlier Sanskrit poets—Mozart employed it to give his piano sonatas a nice sense of balance.[43] It has been (somewhat controversially) argued that Béla Bartók's unusual harmonic system reflects his infatuation with the golden ratio. Bartók once observed that, "We follow nature in composition." Many of his works are broken up into divisions and subdivisions that appear to reflect Fibonacci's numbers.[44]

The most brilliant musical composers are on a par with the greatest mathematicians, especially in the sense that both of their creative solutions often have an air of inexplicability. Because of its great mathematical complexity, Chopin regarded the fugue as "pure logic" in music.[45] Bach was particularly adept in this most difficult form of logic. Allegedly, he improvised for Frederick the Great, king of Prussia, a fugue that proceeded from four parts to five and eventually to a stunning eight parts.[46] Hofstadter's Pulitzer Prize–winning book, *Gödel, Escher, Bach*, goes into fascinating detail on this event. By way of analogy, he crystallized the epic mathematical challenge posed by Bach's fugues: "One could probably liken the task of improvising a six-part fugue to the playing of sixty simultaneous blindfold games of chess and winning them all. To improvise an eight-part fugue is really beyond human capability."[47] Bach's *Well-Tempered Clavier* contains nothing more than a five-part fugue, so it's possible the story is apocryphal. Nonetheless, Bach's mastery of the fugue was unrivaled—to the point that many of his colleagues assumed he had some secret formula to produce such elaborate constructions.[48] Bach was clearly proud of his work and reveled in this reputation. His *Musical Offering* flaunts one inhumanly complex six-part fugue. The only authentic portrait of Bach shows him holding this triple canon for six voices (BWV 1076) as proof of his skill.

FIGURE 33: BACH'S PORTRAIT BY ELIAS GOTTLOB
HAUSSMAN HANGS IN THE UPSTAIRS GALLERY OF THE
ALTES RATHAUS IN LEIPZIG, GERMANY. NOTE HOW HE
PROUDLY HOLDS OUT HIS CANON FOR SIX VOICES.

To appreciate the mathematical implications of this feat, we must first understand how a fugue differs from other types of music. Fugues are a relative to canons, which play a repeating theme against itself. Think of Pachelbel's Canon in D, or *Frère Jacques*, as waves of repeating verses coming one after another. Complexity can be added by changing the pitch or the speed of the repeating verses. Inversion is more complicated: when the original theme goes *up* a certain number of semitones, the melody goes *down* by the same number. Even more complex is a crab canon, a backward copy. In fugues the melodic copies of the main theme are less stringently enforced, and once all the parts have arrived the rules are abandoned. Consequently, fugues possess exponentially greater depth and freedom.

In the *Art of the Fugue*, Bach takes this logic to creative perfection. The whole work is in a single key and employs what appears to be a simple theme. Bach exploits this theme in complex ways, often in four parts, and with increasing complexity and depth as the piece progresses. It culminates, in the final Contrapunctus, with an abrupt ending that seems to be left hanging, incomplete. Eerily, as Hofstadter points out, the final theme ends with the notes B, A, C, and H.[49] (In Bach's Germany, the note B was used for B-flat, and B natural was called H.) The great composer's son, Carl Philipp Emanuel Bach, added a notation to the original manuscript explaining its sudden termination: "In the course of this fugue, at the point where the B.A.C.H. was brought in as countersubject, the composer died."

FIGURE 34: CONTRAPUNCTUS XIV IN THE *ART OF THE FUGUE*, WHEREIN THE LAST FOUR NOTES SPELL OUT B.A.C.H. HIS SON ADDED THE NOTATION AFTER THE COMPOSER DIED.

Bach had been losing his vision for years due to cataracts. It worsened to the point that he finally consented to surgery by the traveling English eye surgeon John Taylor. A recurrence necessitated a second operation a week later. It only made matters worse, and Bach never recovered his vision. Apparently, this was a common outcome from Taylor's work. Though he was extremely knowledgeable about eye diseases, his operating technique left much to be desired. He once admitted that he had blinded hundreds of patients. As one medical historian wryly noted, "This made Taylor a rare combination of a man of serious science and a charlatan in daily practice."[50] Due to his blindness, Bach had to dictate the final prelude of the Chorale to his son-in-law. One day, four months after his surgery, he mysteriously developed a sudden return of vision[xxiii] and, days later, was dead. Hofstadter muses that Bach's death may have resulted from his "attainment of self-reference."[51] The Contrapunctus was left incomplete ... *or was it?*

Undoubtedly, as Wilczek observed in his Nobel speech, Bach was a great manipulator of patterns. Hofstadter, in *Le Ton Beau de Marot*, convincingly argues that Bach does this better than anyone. He builds up a theme, transposes it upward a note or two, and then does it again. Suddenly, just when you expect another go around, "The rug of security would be pulled out from under the listener and a deviating pattern would occur."[52] Hofstadter continues:

> In every intellectual field that I had encountered, ranging from mathematics to music to art to poetry, I had the sense that the moment that patterns were perceived at one level, this immediately established a higher level of abstraction, opening the door to the perception of totally unanticipated types of patterns.[53]

Still Bach inspires awe. His sonatas and partitas are sublime: A single violin interweaves two melodies in perpetual embrace, and double stops—playing two simultaneous notes on different strings—exploited to create an illusion of depth and complexity.

Musical structures are a remarkably good abstraction for the complex patterns found in nature and science. For Einstein, music was not an escape from work but a way to ignite it.

[xxiii] More likely a visual hallucination, like those in Charles Bonnet Syndrome.

Music integrated his math and physics, and enveloped him in the universe's Pythagorean harmonies. Many of Einstein's ideas derived from a music-*esque* wave theory. One of his most famous ideas proved light to be both a wave and a particle. As if making a statement about this theory's connection to music, when the eclipse observations confirmed his calculations, he celebrated with a new violin.[54]

"The scientist does not study nature because it is useful," wrote Poincaré, but rather "because he delights in it, and he delights in it because it is beautiful. If nature were not beautiful, it would not be worth knowing."[55] He was referring not to the superficial beauty of appearances, but to that "profounder beauty which comes from the harmonious order of the parts." Mathematicians often admire the beauty of their own self-made system as if it were itself a work of art, a reflection of God's intentions. The great mathematician Srinivasa Ramanujan taught himself mathematics in an Indian slum. He claimed that his mathematical insights were epiphanies from God. It sounds reminiscent of Stravinsky. For music is also, as composer George Crumb noted, "a system of proportions in the service of a spiritual impulse."[56]

Mathematics is, in many ways, more art than science. Kant wrote that beauty cannot, in fact, be found in mere nature, but only in nature regarded as an analogy to art.[57] In other words, our naïve interpretation of nature as beautiful falsely reassures us of its benign comprehensibility. And so we propose theories that put nature into order. Then, when unexpected connections are revealed, our perception of its beauty is further reinforced. Godfrey Hardy insisted that beauty was the first test of a mathematical proof. "There is no permanent place in the world for ugly mathematics," he wrote.[58] This was a common sentiment among physicists. Paul Dirac and Richard Feynman both made similar statements, as did Hermann Weyl who conceded that, if push came to shove, he would choose the beautiful over the true.[59]

Honeybees could have chosen many different shapes with which to build their honeycomb. Both triangles and squares could equally fill the space without leaving gaps. The ancient Roman mathematician Marcus Terentius Varro's honeycomb conjecture attempted to explain why they instead use hexagons. This six-sided figure, he proposed, furnishes the comb using the least total perimeter, thus conserving valuable wax and labor.[60] Two thousand years later, in 1999, his conjecture was finally proven correct by Thomas Hales.

Sometimes pattern serves a purpose—other times it is purely aesthetic. The bilateral symmetry of a butterfly; the radial symmetry of a three-sided iris, a four-sided jellyfish, or a five-sided starfish; and the hexagonal symmetry of a snowflake all display beautiful patterns amid surrounding clutter. Perhaps we see things as beautiful because of the very fact that it helps simplify nature's complexity. Maybe our imposition of beauty on the world is a necessary adaptation to understanding it. Keats may have said it best when he insisted, in the *Ode on a Grecian Urn*, "Beauty is Truth, Truth Beauty. —That is all / Ye know on Earth, and all ye need to know."[61]

Einstein always appreciated the beauty of simplicity. His assistant Nathan Rosen observed this tautology: "He would aim for simplicity and beauty, and beauty for him was, after all, essentially simplicity."[62] Most scientific endeavors delve into ever more complexity as they progress. But Einstein idealistically sought ever simpler explanations that, he hoped, might culminate in a unified field theory. Einstein compared this quest to Mozart's music, whose pure simplicity magnified its beauty. When he was met with criticism that the universe might not be as beautiful or simple as he hoped,[xxiv] he retorted, "Subtle is the Lord, but malicious He is not."[63]

After Eddington's eclipse results confirmed Einstein's predictions, Planck wrote to Einstein with congratulations, "Thus the intimate union between the beautiful, the true, and the real has again been proved operative."[64] Einstein and Planck shared a love for the juxtaposition of these two virtues. They also both believed that science and religion were complementary. Einstein himself was alternatingly commended, and then castigated, by both sides of the religious coin. Though his theories sometimes contradicted religious scripture, he insisted he was not an atheist. Indeed, he expressed disgust for those exploiting his words in support of their own godlessness.[65] Einstein may not have believed in a *personal* God, but he did have faith in a divinely designed universe. "I believe in Spinoza's God," he professed, "who reveals himself in the lawful harmony of all that exists."[66]

[xxiv] In this case, he was referring to Dayton Miller's so-called proof of the ether's existence. As Einstein expected, it was later disproven.

B♭:

Resonance

Einstein did not grow despondent from his philosophy's repercussions. On the contrary, "One feels the insignificance of the individual," he mused, "and it makes one happy."[1] But others were sometimes crushed under the weight of their beliefs. In the autumn of 1826, John Stuart Mill experienced a profound ennui. His despair grew out of a realization of how meaningless life would be if all his deepest desires were attained. How would he escape boredom? Stuck in Schopenhauer's pendulum,[2] he swung back and forth between struggle and apathy. Music was of great help to him through this period, particularly Carl Maria von Weber's *Oberon*.[3] This brief joy, in a vicious cycle, fomented another round of malaise. Like the philosophers of Laputa who feared the sun would burn out, Mill became tormented by the idea that music could also become exhausted. At some point, he lamented, the succession of great composers must come to an end: "The octave consists only of five tones and two semitones … [These] can be put together in only a limited number of ways, of which only a small proportion are beautiful: most of these, it seemed to me, must have been already discovered."[4] What then?

Mill's emotion originated in the primitive depths of his brain, in the amygdala. Within the amygdala is a small collection of cells specifically responsible for our emotional response to music. This is the nucleus accumbens. It ignores randomly played tones but responds powerfully to musical patterns. Many of the great thinkers in this book capitalized on this feature of the mind.

Einstein frequently struggled with communicating emotion. With his scientific colleagues, it gave him an aura of aloofness. It led to the unraveling of his first marriage to Mileva Marić. It constructed barriers between him and his children as well. Einstein and his son Eduard eventually found emotional resonance in music. Playing violin together was the only way they could meaningfully communicate.[5] Educator, pianist, and child whisperer Mr. Rogers also endorsed the ability of music to touch the deepest recesses of our spirit. "Music was my first language," he wrote. "I could literally laugh or cry or be very angry through the end of my fingers."[6] What Kafka said of literature is also true for music: "[It] must be the axe for the frozen sea inside us."[7]

The emotional power of music can exploit or inspire the masses. The passion excited by a military march would hardly be feasible without the music. Every commercial has a jingle, every movie a soundtrack. Music's range far surpasses nearly any other tool's power for influencing emotional state. This led the early Catholic church to ban polyphonic music out of fear that it might stimulate arousal. They feared it could even incite doubt about the unity of God. The church incriminated the tritone as *diabolus in musica*, so dissonant that it must be the work of the devil. They held the same view of certain unmentionable numbers, still called *irrational*, because they have no pleasing ratio or proportionality. The tritone interval did not gain acceptance in musical compositions until the Baroque period. Modern composers now use them with abandon.

Humans are gregarious animals, and music may have evolved as a social adhesive. It can arouse, it can soothe. It can inspire joy or, alternatively, lugubriousness. British musicologist Charles Avison described the sheer variety of emotions that music can stir. "By the musician's art," he wrote, "we are by turns elated with joy, or sunk in pleasing sorrow, roused to courage, or quelled by grateful terrors, melted into pity, tenderness, and love, or transported to the regions of bliss, in an extacy [sic] of divine praise."[8]

From an evolutionary standpoint, an adaptation must improve mating success if it is to be propagated. Music's ability to express emotion may be part of this equation. If Shakespeare was correct, and music is the food of love, it must offer some reproductive benefit. It certainly played to Einstein's advantage: his second wife, Elsa, once confided that she fell in love with Einstein "because he played Mozart so beautifully on the violin."[9] Evolutionary psychologist Geoffrey Miller argued that music may be a marker for creativity and problem-solving.[10] Others have argued that our emotional attachment to music is just an evolutionary parasite—an unexpected byproduct of natural selection.[11] Although musical ability may have initially been promulgated for utilitarian purposes, this theory goes, now it just pulls at our heart strings. Exaptation occurs when a trait that originally evolved for one reason is later repurposed. The swim bladder of fish, for example, evolved into the mammalian lung. Peacock feathers morphed into a courtship display that has nothing to do with flight or insulation. Whatever the reason that humanity went all-in on music, it must have been an important driver of reproductive success, as its influence and importance continue to this day.

FIGURE 35: EINSTEIN AND HIS WIFE, ELSA, EARLY 1920s.

The overlap between music and emotion are particularly highlighted by disorders that affect social and emotional functioning. Patients with autism, for example, struggle with social interactions. They also frequently "don't get" music.[12] Conversely, those with Williams syndrome are both extraordinarily social and musical, despite very low cognitive abilities. The juxtaposition of autism and Williams syndrome emphasizes the interrelationship of music and gregariousness. Emotional intelligence may be linked to music in ways we don't understand. Certainly, the two seem to employ similar mental frameworks.

Music has a tremendous influence on our emotional interaction with the world. It exerts these effects through a deluge of dopamine and other neurotransmitters.[13] Pattern recognition is so important to our survival that our brains reward it with dopamine. It is like a parent reinforcing good behavior. Dopamine is responsible for the musician's sense of aesthetic rapture and the true believer's religious rapture.

It is the brain's primary reward chemical, and it takes its job seriously.

How powerful is it? Consider that cocaine, the most addictive drug on the planet, primarily acts by increasing dopamine.[14] Dopamine is so alluring that we sometimes even seek it at our own peril. Addiction is directly related to the effects of dopamine on the reward center of the brain. Drugs that artificially raise dopamine levels in the brain, like some medications for Parkinson's disease, can even trigger unusual addictions like gambling. Tellingly, the nucleus accumbens, the focal point of musical emotion, is also the seat of addiction's reward center. Pathology versus functionality is often just a matter of degree.

Depression and anxiety also involve many of these same neurotransmitters. Author William Styron once suffered from a paralyzing depression. He disliked that term, though, claiming it was "a true wimp of a word for such a major illness."[15] As his despair progressed beyond all hope, he couldn't even find the energy to write a suicide note:

> But even a few words came to seem to me too longwinded, and I tore up all my efforts, resolving to go out in silence ... My wife had gone to bed, and I had forced myself to watch the tape of a movie in which a young actress, who had been in a play of mine, was cast in a small part. At one point in the film, which was set in late–nineteenth-century Boston, the characters moved down the hallway of a music conservatory, beyond the walls of which, from unseen musicians, came a contralto voice, a sudden soaring passage from the Brahms Alto Rhapsody.

> This sound, which like all music—indeed, like all pleasure—I had been numbly unresponsive to for months, pierced my heart like a dagger, and in a flood of swift recollection I thought of all the joys the house had known ... All this I realized was more than I could ever abandon, even as what I had set out so deliberately to do was more than I could inflict on those memories, and upon those, so close to me, with whom the memories were bound.[16]

Styron's intensely vivid account illustrates music's deep impact upon the soul. There are not many other things—perhaps no other thing—that could bear such a great force upon the psyche at a time when the conscious mind has already given up the ghost. Schopenhauer expressed this when he wrote, "The inexpressible depth of music, so easy to understand and yet so inexplicable, is due to the fact that it reproduces all the emotions of our innermost being, but entirely without reality and remote from its pain."[17]

Galileo also turned to music during trying times. After publicly supporting the Copernican theory of heliocentrism, his *Dialogue Concerning the Two Chief World Systems* managed to both contradict Holy Scripture and insult Pope Urban VIII. He was charged with heresy and placed under house arrest for the remainder of his life. Isolated and forlorn, his beloved lute was the only thing that still sparked joy. In passionate letters, he detailed how his senses became drunk on its mellifluous juices: "The effect of the fifth is to produce a tickling of the eardrum, so that its gentleness is modified by sprightliness, giving the impression simultaneously of a gentle kiss and of a bite."[18]

After Max Planck lost his son Karl in World War I and both twin daughters to childbirth, he admitted to his colleague Hendrik Lorentz, "There have been times when I doubted the value of life itself."[19] It's unlikely many of us would be able to persevere through such profound devastation. But Planck found succor in music. "In the weeks and months following Emma's passing, solace sometimes arrived at the Planck home carrying a violin case," wrote his biographer Brandon Brown. "The trio of Einstein, Erwin, and Max Planck mourned within their music."[20] Later, after his son was executed for plotting against Hitler, Planck grieved by playing Erwin's favorite songs on the piano. Eventually, he realized that unhappiness was not inevitable. His scientific work called him: "There are still many precious things on the earth and many high callings, and the value of life in the last analysis is determined by the way it is lived."[21] Oliver Sacks, who also suffered from mental illness, described this healing effect of music. "There is a tendency in philosophy to separate the mind, the intellectual apparatus, from the passions, the emotions ... Yet music calls to both parts of our nature—it is essentially emotional, as it is essentially intellectual."[22]

Watson contemplated this rare quality of music as well. In *The Valley of Fear*, he was surprised to discover that music even affected the criminal mind. "Strange indeed is human nature. Here were these men to whom murder was familiar, who again and again had struck down the father of the family ... and yet the tender or pathetic in music could move them to tears."[23] Hans Albert Einstein told how his father often played the violin in a vain attempt to quiet the kids.[24] Holmes did the same thing for Watson in *The Sign of Four*.

> "Look here, Watson; you look regularly done. Lie down there on the sofa, and see if I can put you to sleep." He took up his violin from the corner, and as I stretched myself out he began to play some low, dreamy, melodious air—his own, no doubt, for he had a remarkable gift for improvisation ... I seemed to be floated

peacefully away upon a soft sea of sound, until I found myself in dreamland.[25]

Music was also therapeutic for Nietzsche. Merely discussing his feelings just wouldn't cut it. "Compared to music," he wrote in *The Will to Power*, "all communication by words is shameless. Words dilute and brutalize; words depersonalize; words make the uncommon common." His friend, Richard Wagner, empathized with him on this. He understood that music transcends the individual to express that which is "inexpressible to any other language."[26] The emotional power of music eclipses language and culture—and maybe even species. Orpheus and his lyre were no mere fable.

F:
Synesthesia

EVEN AFTER THE ONSET OF BEETHOVEN'S DEAFNESS, HE CONTINUED COMPOSING music in his head as he imagined its sound. Dmitri Shostakovich really could hear music in his head. During the German siege of Leningrad, he suffered a shrapnel injury to the auditory cortex of his brain. He was reluctant to have it removed, and for good reason: It endowed him with an unusual skill. If he tilted his head just right, he would hear music. As he straightened up again, the music stopped.[1] He made use of this strange superpower, exploiting bits and pieces of hallucinated music in his compositions.

Many other composers have been blessed, or cursed, with musical hallucinations. As a child, Tchaikovsky's parents once found him whimpering in bed. "This music!" he cried. "It is here in my head. Save me from it!"[2] And Schumann, whose musical hallucinations may have been a prodrome of neurosyphilis, imagined "wondrously beautiful pieces of music, fully formed and complete."[3] Eventually, Schumann's beautiful musical hallucinations became full-blown schizophrenic delusions. He was harassed, day and night, by a ceaseless rushing of music that gave him no peace. "Oh! how near are genius and madness," wrote French philosopher Denis Diderot. And how fascinating that music is so frequently intercalated between the two.

Yet other conditions further elucidate the myriad ways music insinuates itself into the most remote parts of the brain. Recall that color and sound exist along the same continuum of wavelength. Occasionally, these sensory systems become jumbled during development. This strange cross-activation is surprisingly common in newborns but usually disappears as the brain matures.[4] About one in two thousand people, though, never outgrow it. In these cases, the brain misinterprets one type of signal for the other. This is synesthesia. Some see different colors with different numbers. Feynman probably had this type of synesthesia. "When I see equations, I see the letters in colors—I don't know why ... I see vague pictures of Bessel functions ... with light-tan *J*'s, slightly violet-bluish *n*'s, and dark brown *x*'s flying around. I wonder what it must look like to the students."[5] Others see colors with particular words or letters. Wittgenstein envisioned that "the vowel e is yellow," for example.[6] Some even experience tastes rather than seeing colors. In all, over 80 different types of synesthesia have been described.

In the musical variety, affected individuals see colors when they hear music. Different notes or keys spawn different colors, and timbre influences color saturation.[7] They are consistent, though idiosyncratic; each affected person has different color/tone matchups. For instance, Oliver Sacks wrote about the composer and synesthete Richard Torke, who referred to G minor as the key of ochre. G major, on the other hand, was bright yellow.[8] Because of this precise color specificity, Torke considered it inconceivable that a synesthete would not also have perfect pitch. It is not clear whether synesthesia is more common in musicians, but they are certainly more aware of its presence. Rimsky-Korsakoff and Messiaen both had synesthesia. Scriabin and Ravel probably did too. These composers referred to music as "sound paintings."[9]

Ernst Chladni, in the late eighteenth century, made sound visible to the rest of us. A Hungarian lawyer, his interests later turned to physics and mathematics. Chladni placed sand on differently shaped metal plates and used a violin bow to create vibrations on the plates.[10] The sand oriented in lines representing the sound waves. The resulting images revealed a beautiful hidden world of sound. Modern scientists and artists have adapted Chladni's techniques using modern speakers to create even more stirring images.[11] Sand accumulates at nodal points where the standing wave has an amplitude of zero (that is, where the plate is not moving). Here, the sand forms lines that correspond to the wavelength of the pitch [figure 36]. As the tone shifts, the sand spreads out until the resonance frequency of a pure note is reached, and then, suddenly, the sand jumps into a new configuration—just like a quantum leap! With shorter wavelengths, high frequency tones create vastly more complex patterns.

Figure 36: Ernst Chladni's original publication depicted the designs seen with different musical notes.

Whereas many synesthetes are also aesthetes, other conditions lead to opposing effects. Some people, try as they might, just do not appreciate music at all. Freud, as discussed earlier, was distinctly unappreciative of its charms. Intriguingly, his colleague Carl Jung had a similar perversion. In Jung's autobiography, he compared a "singing" kettle to music, "which in reality I cannot abide."[12] It's not that he didn't recognize music's relevance. In fact, he liked the idea of incorporating music into traditional psychoanalysis. "Music should be an essential part of every analysis," he told concert-pianist-turned-music-therapist Margaret Tilly, because it "reaches deep archetypal material that we can only sometimes reach in our analytical work with patients."[13] But despite his intellectual appreciation of music's benefits, he couldn't *enjoy* it. He told Tilly, "My mother was a fine singer, so was her sister, and my daughter is a fine pianist. I know the whole literature—I have heard everything and all the great performers, but I never listen to music anymore. *It exhausts and irritates me* [emphasis added]."[14]

True amusia is so atypical that it is considered pathologic. In the case of these two psychoanalysts, it is tempting to attribute their dislike for music to an excessively logical mind. The right inferior frontal gyrus encodes musical pitch, and this area is demonstrably underdeveloped in the amusic brain.[15] There are also reports of patients with damage to the superior temporal gyrus (that of the aha moment) developing amusia.[16] It would be fascinating to know whether Freud and Jung had either of these anatomic anomalies. Freud may have had true amusia, or even melophobia; but Jung, in contrast, was probably just too emotionally attached to it. One neurological review proposed that the break between the two psychoanalysts occurred because Freud saw the world through a left-brained lens, while Jung saw it through a right-brained one.[17] As Tilly surmised, "[Jung] cares too much, not too little."[18]

Vladimir Nabokov presents an unusual case because he

combined synesthesia with a dislike of music. He was a verbal synesthete: he saw different colors with different words or phonemes. He described the letter *h* as a "drab shoelace," while the letter *a* was reminiscent of "weathered wood."[19] Perhaps these images facilitated his lyrical writing style. The juxtaposition with amusia is particularly unusual. In his memoir, he wrote, "Music, I regret to say, affects me merely as an arbitrary succession of more or less irritating sounds."[20] It is hard to fathom someone so artistic and lyrical in one realm, yet so mundane and unimaginative in another. Clearly, the patterns that drive music and language are vastly more complex than we imagine.

Other conditions provide additional insights on the role of music in the brain. German writer Novalis[xxv] could have been thinking about Parkinson's disease when he wrote, "Every disease is a musical problem. Every cure is a musical solution."[21] Music therapy is now an established treatment modality for patients with Parkinson's disease. One of the hallmarks of this disorder is a difficulty with smooth movements, especially the start-stop variety. Simply walking, for instance, has a gait cycle more analogous to a grandfather clock's swinging pendulum than to a rolling wheel. Tricking the mind into using different regions of the brain to perform the task can produce startling results. Patients who can barely walk suddenly dance fluidly. They quickly reach out to catch a thrown ball. They ride a bike: A video from the *New England Journal of Medicine* shows an elderly gentleman struggling to walk.[22] He has the typical slow movements, shuffling steps, and festinating gait of advanced Parkinson's disease. But once his therapist helps him onto a bicycle, he easily cruises around the parking lot, spryly hopping off with a flourish at the end. The smooth continuous motion of pedaling a bicycle is more reminiscent of its rolling wheels than of walking's pendulum. It is legato, not staccato. Sacks described

---

[xxv] With a name like Georg Philipp Friedrich Freiherr von Hardenberg, it's no wonder he used a *nom de plume*.

this phenomenon in the concert pianist Lukas Foss. Suffering from Parkinson's disease, Foss "rocketed" uncontrollably to the piano where he proceeded to play a beautiful and complex Chopin nocturne.[23] Music seems to bring a different tempo and rhythm to even routine activities.

Think back to the ancient kinship between language and music that we discussed earlier. It is on remarkable display in certain language disorders—like dyslexia, for example. This common deficit affects about 10 percent of children. These kids (and adults) grapple with spelling, reading, and writing, despite otherwise normal intelligence.[24] The primary deficit is, unintuitively, a phonological one: it is the individual phonemes and syllables in a word that cause difficulty. Mounting evidence points to a deficit in processing dynamic auditory information.[25] The surprising connection of dyslexia to rhythm is succinctly summarized by researcher Katie Overy:

> It appears that dyslexics' difficulties with music may be closely connected with their language and literacy difficulties, and that all of these difficulties are based upon fundamental timing problems. It therefore seems reasonable to propose that a music-training programme which focuses on rhythm and timing skills may develop particular cognitive skills involved in language processing, and serve to remediate some of the difficulties which dyslexics experience.[26]

The deficit in dyslexia is more of a patterning and networking problem than one of "atheoretical cartography." There are, however, a few abnormalities that can be mapped to specific locations in the brain. Interestingly, one of these is in the left inferior frontal gyrus—the same structure that is shrunken on the right side in patients with amusia. We have already seen that music builds greater interhemispheric communication and, moreover, that there is a strong correlation between music and language. It should not be surprising, then, that research shows musical training can successfully improve reading skills in dyslexic children.[27] Building connectivity between the left and right hemispheres can help bypass the deficits of dyslexia.[28] What is surprising is that this tool is not utilized better.

For a remarkable proof of the brain's plasticity, consider Rasmussen syndrome. These children have severe seizures that, oddly, only affect half the brain. An early treatment for this devastating condition was a surgical hemispherectomy. Vastly more aggressive than Roger Sperry's procedure (which severed the communication between the two halves of the brain), a hemispherectomy removes fully half of the brain. Remarkably, those with a left hemispherectomy, who should by all rights lose their language ability, somehow retain normal language skills.[29] The right side picks up the slack to become the *de facto* language center.

Figure 37.

Singing and speech, although related, are controlled by different areas within the brain. Researchers have ingeniously proven this by selectively anesthetizing just half of the brain. When only the left hemisphere was sedated, subjects could not speak—but could still sing![30] The opposite effect occurred when the right side was sedated. Keep in mind, though, this delineation is not as discrete as implied by these findings. The singing side of the brain is also recruited for certain speech tasks, employing the right side's broad pattern recognition skills when appropriate. It is important for understanding metaphor, for example, including vague inferences and jokes.[31] Patients with damage to the right brain, who seemingly possess normal speech, often miss the gist of conversations when the point is not explicit. This is the aptly named Amelia Bedelia syndrome.

The brain's plasticity is not nearly as great in adults as it is in children. It is by no means absent, though. A language deficit that often occurs in adults after a stroke, called aphasia, results from damage to the language centers of the brain. If the damage specifically strikes Broca's area in the posterior frontal lobe, patients are left with expressive aphasia—also known as Broca's aphasia—and can speak only haltingly, if at all. One patient with this condition, described more than 250 years ago, had speech so limited he could only say "yes." Despite this, he could still sing religious hymns beautifully.[32]

Jill Taylor wrote the remarkable book *My Stroke of Insight* about her experience as a stroke survivor. She suffered a large hemorrhagic stroke in her left brain. Because she was right-handed, this was likely her dominant hemisphere. As a neurobiologist, she knew exactly what the implications of this were. As recognition set in, she thought, "Oh my gosh, I'm having a stroke!" But that fear soon yielded to the scientist's curiosity: "Wow, this is so cool!"[33] Prior to the stroke, she had played cello and guitar. Perhaps this contributed to her eventually complete recovery. A surfeit of connectivity within the musician's brain, we now know, markedly improves its plasticity and cognitive reserve. As for Dr. Taylor, she has continued singing and playing her guitar, probably, she notes, with even more joy than before. "When I am simply grateful," she writes, "life is simply great."[34]

Just like with creativity, getting the two halves of the brain to work together can overcome language difficulties. For these adult patients with aphasia, this is particularly true. Patients with chronic aphasia after a stroke are notoriously resistant to further progress. Most improvement is made in the days and weeks immediately after a stroke. One study employed music therapy as a complement to speech therapy over fifteen weeks in patients who had their strokes an average of almost four years prior. The music therapy patients showed a stunning 50 percent improvement, while those receiving speech therapy alone experienced no change. Interestingly, just like in dyslexic children, the rhythm component of therapy was the most effective. The authors surmised that, "In view of the similarity/overlap of the music and speech areas in the brain, rhythm may organize/synchronize actions." [35] In other words, music acts like an orchestral conductor for the language parts of the brain.

We often don't really miss something until it's gone. Dr. Sperry's experiments did this for defining the specialized roles of the two cerebral hemispheres. We've come to see the left brain as focused and discriminating and the right side as fuzzier and more expansive, seeing the big picture. Science often progresses from the study of disease patterns. Music's influence on many common medical conditions highlights another profound truth: the ability to function in our daily lives is deeply and deceptively reliant upon adaptations that originally evolved to facilitate musicality. Consequently, anomalies in the complex latticework of the musical brain impact us in numerous, often unforeseen ways. Many diseases are indeed musical problems, and every developmental misstep instructs us in a new and fascinating way—sometimes even with a musical solution.

C:

Apophany

WE HAVE RETURNED ONCE AGAIN TO THE TONIC, BACK TO THE C. THIS IS AS good a time as any to point out just how much sleight of hand was needed to end up exactly here. That is because the circle of fifths, as I hinted at the beginning, is not as simple as it appears. Recall how Pythagoras discovered that increasing by an octave was accomplished by halving the string length—a 2:1 ratio. And fifths, remember, increase by a 3:2 ratio. We are now faced with the incontrovertible fact that here our math breaks down. As musician and mathematician Gareth Loy asserted, "Contrary to the wishes of scale builders and musicians from antiquity to the present, the powers of the integer ratios 3/2 and 2/1 do not form a closed system."[1] A well-established mathematical proof shows that no power of two can ever equal any power of three. The twain shall never meet. Sequential perfect fifths will never return to the same note as the octave will. Music theorists fudge this with a correction factor, called the Pythagorean comma, in order to prevent our circle of fifths from becoming a spiral of fifths.

Let me explain: We begin on the lowest key on the piano, A_0, at 27.5 Hz. Increasing by seven octaves, each a 2:1 ratio (so doubling every step), our math proceeds: **27.5**, 55, 110, 220, 440, 880, 1,760, **3,520**. If, instead, we increase by fifths, at a 3:2 ratio, the numbers now proceed: **27.5**, 41.3, 61.9, 92.8, 139.2, 208.8, 313.2, 469.9, 704.8, 1,057.2, 1,585.8, 2,378.7, **3,568**. By the time the fifths come back around, at A_7, they are now ringing in at 3,568 Hz, not 3,520. The difference between the math and our naming convention defines the Pythagorean comma: a factor of 1.0136..., or 23.46 cents in the logarithmic units used by musicians.[2] Pythagoras, by the way, would roll over in his grave if he found out an irrational number was named after him.

 Musicians deal with this mathematical problem with a tuning trick called temperament, which compromises the purity of the fifths in order to keep the octave pure.[3] Enharmonic equivalents are notes that are considered the same but have different names. For instance, B-sharp and C are the same note on a piano keyboard, as are G-flat and F-sharp. Modern Western music calls this *equal temperament*. The Pythagorean comma makes up for the fact that G-flat and F-sharp are not *mathematically* the same note. Equal temperament demands we slightly detune everything to spread out the error over many intervals so that each one still sounds reasonably good. American composer Terry Riley thinks equal temperament is a travesty. He once proclaimed that "Western music is fast because it's not in tune."[4]

 Bach, as we've seen, was not averse to musical challenges. His *Well-Tempered Clavier* was an attempt to write a cycle in every key. It was a great mystery, at the time, whether his E-flat minor Prelude would be in tune with the D-sharp minor Fugue.[5] Well-tempered and equal tempered, by the way, are not the same. Numerous other tuning systems were devised prior to the general acceptance of equal temperament. In the Renaissance, for example, this tempering was unequally divided across the octave, so different intervals had their own color and dissonances. Some of Bach's contemporaries wrote in

Werckmeister III, while others composed in the temperament ordinaire, Kirnberger temperament, or septimal meantone, to name just a few. Because different temperaments change the sound of certain thirds and fifths, it's likely many compositions were intended to sound quite differently than we hear them today in equal temperament. A group of music scholars attempting to reproduce some work by Orlande de Lassus struggled mightily with this issue. Finally, "Bingo! – the eureka realisation that in Lassus's world quarter-comma meantone temperament was commonplace."[6] Once they sharpened the fourths, their tuning problems resolved.

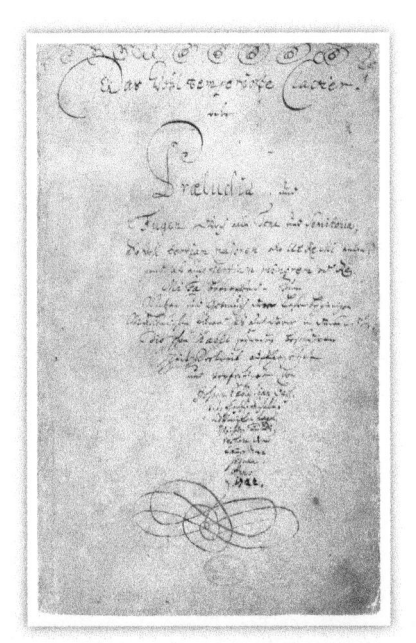

FIGURE 38: THE TITLE PAGE OF BACH'S *WELL-TEMPERED CLAVIER* SHOWING HIS CURLICUED TEMPERAMENT MARKINGS AT THE TOP.

In the *Well-Tempered Clavier*, Bach likely wrote in a unique temperament. Mysteriously, we still don't fully appreciate what it was. We are likely hearing Bach's compositions—and still loving them—in a temperament that doesn't fully capture their original complexity. The scrolling loops underlining each chapter heading in this book are taken from the title page of his 1722 composition of the *Well-Tempered Clavier*. The meaning of these loops has puzzled musicologists for ages. Several attempts to reconstruct Bach's temperament have largely failed. In 2005, musicologist Bradley Lehman came up with a breakthrough idea.[7] After noticing the scrollwork contained eleven loops of three different types (simple, double, and convoluted), he had the epiphany that this could describe the temperament of the twelve fifths (in the circle of fifths) if the first note was known. He identified a hidden C attached to the last loop on the right, likely indicating the starting note. He suggested the convoluted spirals represent one-sixth of a Pythagorean comma, the double loops are one-twelfth, and the simple loops are pure fifths. With this as his starting point, Lehman devised a temperament that "brings out qualities of Bach's composition that are hidden in equal temperament."[8] Subsequent researchers pointed out discrepancies with Lehman's original idea. Some thought it worked better if turned upside down or if the squiggles meant different things. Musicologists now generally agree that it is a marking for temperament, but its precise interpretation remains uncertain.

Of course, there is another infinite world of pitches between the keys of the piano. Composer Harry Partch once attempted to bridge this divide by cutting the octave into forty-three microtones derived from the natural harmonic series. But, as Powers pointed out in *Orfeo*, forty-three pitches don't get you any closer to infinity than twelve.[9] My point in entertaining this esoteric diversion is that composers have had to creatively rectify these discrepancies ever since Pythagoras first defined the octave and the perfect fifth. It is relevant to music's relationship with math, and to how both relate to the natural world.

Mathematical proofs are extremely precise when they describe other mathematical constructs. The most renowned mathematical theorems fall into this category. Consider the prime number theorem, the irrationality of the square root of two, and Gödel's incompleteness theorem. We know what we're talking about—when we know what we're talking about. But when mathematics tries to describe the natural world, its precision smudges a bit. Even the greatest minds resort to fudge factors to make the math work. Schrödinger's equation does a great job at modeling a hydrogen atom, but it can't do the same for helium. Its genius has been described as more of an invention than a discovery, "an arbitrary, contingent, historical accident" that matches real world data only with the help of "hideously complex patches and approximations," as science journalist John Horgan put it.[10]

These are mathematical *models* in the real sense of the word: a reminder that the map is not the territory. Poincaré might have described the Earth's sphere as topologically different from the flat plane of a map. Tobias Dantzig, on the other hand, delightfully brought it back to music: "The harmony of the universe knows only one musical form—the legato; while the symphony of numbers knows only its opposite—the staccato. All attempts to reconcile this discrepancy are based on the hope that an accelerated staccato may appear to our senses as legato."[11] We ought to remind ourselves that this indefatigable quest for continuity—for legato—is why "math is not so much a science as an art."[12] The world, it seems, is not as majestically simple as we've been led to believe. "The hallmark of most natural phenomena," John Barrow wrote, "is a deep complexity masquerading as simplicity."[13]

The question again arises: Are our perceptions of pattern and symmetry epiphany or *apophany*? Consider the circle, the most perfectly symmetrical shape, displaying rotational symmetry in every direction. da Vinci's *Vitruvian Man* stands inside a circle to demonstrate the ideal proportions of his body. Certainly, God would design the universe with this same degree of perfection. This very bias led Kepler to spend years beating his head against the data, trying to make it fit, before realizing that Ptolemy's vision of circular orbits was wrong.[xxvi] Ironically, and perhaps tellingly, it was his very belief in this beautiful simplicity that guided him to an alternative solution. Ellipses, he rationalized, are like circles in that both are created by slicing a cone in different planes.

Aristotle described the four earthly elements as earth, fire, water, and air. He reserved a fifth element, the ether, for the heavens. Einstein's predecessors wasted years of scientific investigation explaining phenomena in the ether, which enabled their beautiful wave theories of light to work. Scientists performed all sorts of intellectual acrobatics to make their science fit this ideal. Finally, Michelson and Morley proved it did not exist, paving the way for Einstein's quantum theory and special theory of relativity.[xxvii] Light, he convincingly argued, wasn't always a wave. Like the mind-matter duality of Descartes, it could exist as both a wave and a particle at the same time. This allowed light to travel through the vacuum of space without the benefit of ether.

Philosopher Karl Popper, himself a pianist, was not against arguments for beauty in science, but he wanted us to look at the simplest, most beautiful theories first only because they are the most easily testable.[14] Philosopher Thomas Kuhn also wrote that beauty, in the sciences, is, at best, a tool: "A criterion of choice between theories that are in other respects comparable, or a guide to the imagination seeking a key to

---

[xxvi] Interestingly, our neighboring planet Venus does, in fact, have an almost perfectly circular orbit. (Clarke-Ezzidio, "Venus Isn't Habitable.")

[xxvii] The optical instruments used by Michelson and Morley in their experiments were invented by Karl Rudolph Koenig. Koenig, a student of Helmholtz, was a talented violinist and later a violin luthier. (Root-Bernstein, "Music, Creativity and Scientific Thinking," 63–68.)

the solution of an intractable technical puzzle."[15] String theory presents another example of where our infatuation with beauty may deceive us. One of its chief proponents, Edward Witten, speaking with John Horgan at *Scientific American*, highlighted string theory's "incredible consistency, remarkable elegance and beauty."[16] The problem with string theory is not its beauty. It is that after more than half a century of theorizing, there is still no proof of its accuracy. Indeed, it—and its predictions of alternative dimensions and multiverses—are neither testable nor falsifiable. This places it squarely in the company of Popper's pseudoscience. When beauty is itself the end, it begins to mislead.

Benoit Mandelbrot's fractals have found some purchase in defining the chaotic patterns of nature. The term itself derives from the Latin *fractus* and denotes something irregular and fragmented—like a broken-up stone. "With two hands," Mandelbrot observed, "you can count all the simple shapes of nature. Everything else is rough."[17] In the *Fractal Geometry of Nature* he provided examples: "Clouds are not spheres, mountains are not cones, coastlines are not circles, and bark is not smooth, nor does lightning travel in a straight line." Intriguingly, the very mathematics designed to display nature's innate disorder, is itself a paragon of beauty [recall figure 30].

Rothstein feels certain that science's infatuation with mathematics will persist. "The faith is not just that mathematics represents the truth," he wrote, "but that it represents the truth so much more faithfully than reason gives us any right to expect."[18] It was self-evident to Einstein that "Nature is the realization of the simplest conceivable mathematical ideas."[19] This was an echo from the past, from Newton, whose giant shoulders Einstein stood upon. In the third book of his *Principia*, he wrote, "Nature is pleased with simplicity." It gave Einstein pause on the occasions when he considered the deeper implications of this philosophy. He was fascinated by the fact that mathematics, a product of human thought, so admirably reflected reality. In a lecture to the Prussian Academy of Sciences, he expressed genuine faith that human reason alone could decipher profound universal truths.[20] His fanatical dedication to this philosophy may have condemned him to failure with his loftiest goal. A unified field theory—a "theory of everything"—continued to evade even the great Einstein's own reasoning.

The scientific principles designated as the so-called laws of nature are not divinely decreed. Rather, they are post hoc correlations of an observed pattern within the greater tapestry of the universe. With our Procrustean saw—the irrational constant—we methodically cut our guests to fit their beds. The number of constants in regular use are mind-boggling. The standard model of particle physics requires at least nineteen of them, and the standard cosmological model adds at least another twelve.[21] Many of these are irrational. To just name a few: The imaginary number (i) represents the square root of -1; Euler's number (e) is approximately 2.71828... ; phi (φ), the golden ratio, is 1.61803... ; and pi (ϖ) is of course 3.14159... .[xxviii] Planck's constant (h) is not technically an irrational constant, as it can be calculated to a precise $6.62607004 \times 10^{-34}$. Schrödinger exploited two of these (h and i) when he devised his famous equation, and Euler's identity makes use of three: $e^{\varpi i} + 1 = 0$. The Pythagorean comma brings the same faux precision to the musical side of mathematics.

The spiral of fifths is evocative of the spirals generated by Fibonacci's sequence. Hardy's and Feynman's beautiful

---

[xxviii] As of the time of this writing, pi has been calculated to more than sixty-two trillion places after the decimal point! (Delbert, "Supercomputer.") Astronomer Simon Newcomb points out that "Ten decimals are sufficient to give the circumference of the Earth to the fraction of an inch, and thirty decimals would give the circumference of the whole visible universe to a quantity imperceptible to the most powerful microscope." (Dantzig, *Number*, 121) Clearly, our desire to calculate pi to fifty trillion places is not driven by utility but by curiosity and ambition.

mathematics do not always flawlessly represent the real world's chaos and disorder. But its patterns—irrational constants be damned!—make the universe at least partially intelligible to our earthbound minds. We have beautifully simplified its disorder into an intelligible and consistent system. Few appreciate that nature is not as discretely and tidily arranged as we make it out to be. For one, there is no actual partition between biology, chemistry, and physics. Do we really think nature treats them as discrete, operating under different laws and rules? We have merely created artificial boundaries to aid our own primitive understanding.

This may at least partly explain why mathematics and music are so exquisitely intertwined and why they seem so useful. Music is not *really* mathematics. They just share certain characteristics that model the abstractions inherent in nature. In so doing, they elucidate a mental model our mortal minds can exploit to gain a sense of divine understanding. "We must learn a new modesty," wrote Hermann Weyl. "We have stormed the heavens, but succeeded only in building fog upon fog, a mist which will not support anybody who earnestly desires to stand upon it."[22]

Figure 39.

Lucretius's atoms, from the Latin *atomos,* were supposed to be "uncuttable." But four hundred years earlier, Anaxagoras had already perceived that "among the small there is no smallest, but always something smaller."[23] First we identified protons, neutrons, and electrons within the supposedly uncuttable atom. Now physicists at the Large Hadron Collider, again proving Anaxagoras correct, smash atoms to pieces at speeds that would have them circling the Earth seven and a half times per second. These fulminations reveal not just hadrons, but quarks, leptons, mesons, pions, and bosons. Deeper still (finally?) are the gluons, photons, higgs, taus, muons, and up, down, top, bottom, charm, and even strange particles. Nature more and more resembles an infinite Mandelbrotian pattern, revealing ever more complexity the closer we look. There is always something smaller, and there is always something bigger.

"Whoever undertakes to set himself up as a judge of Truth and Knowledge," wrote Einstein, "is shipwrecked by the laughter of the gods."[24] We display great hubris when we fall prey to believing we truly comprehend, or even worse control, nature's patterns. As Horgan said, we must "counter our terrible tendency toward certitude."[25] We run the risk of causing ourselves and our world great harm when we naïvely overestimate our understanding.

We close, then, with a humble appreciation of our impotence in the face of nature's vast and complicated framework. This, Einstein noted, was Spinoza's God: a majestic empyrean system, not a personal God attentive to our needs. Music and math are simply metaphors for the world. Like all metaphors, they lubricate the pathways to insight. These metaphors, though, have such great complexity that they can surpass even their own attempts at explication.

FINALE:
Sagacity

Musicologist Anthony Storr wrote that our "stream of consciousness" is more like a "stream of unconsciousness."[1] We have far less free will than we believe and often make decisions on autopilot without real awareness. Daniel Kahneman called this "thinking fast."[2] It is instinct, not the deliberate weighing of decisions that occurs in slow thought. Consciousness is but a thin veneer—twigs floating on the stream's surface. The more critically we learn to think with the left brain, the more we can spot our own biases and recognize where we may be led astray. But excessive logic leaves no room for originality. So, we must also learn to think with the right brain. Not only to assist creative problem-solving but also to bear witness to the beauty inherent in the dichotomy of simplicity and complexity all around us. Most importantly, we must get these two halves working synergistically. This is where music can have its most dramatic impact.

It is staggering, after all this, that music is playing less and less of a role in our educational system. In some quarters it is even considered an extravagance—dispensed with altogether. It seems evident that our children need to learn *how* to think more than they need more rote learning. Music, as Plato and Boethius argued so many years ago, is an integral part of active learning. Like canaries who grow extra brain cells to help them compose more complex songs, our children also need the stimulation of music to attain optimal brain development. The modern-day equivalent to the quadrivium is the four c's: critical thinking, creativity, collaboration, and communication. Let's not forget the role of music in all four aspects of this curriculum.

It should now be readily apparent how deeply music's evolutionary taproot extends. What we think of as language would likely not exist without it. Harmony was our first syntax and melody our first grammar. Music transcends the weariness of Schopenhauer's pendulum to provide social and cultural solace: *Gemeinschaft and Gesellschaft*. Musical compositions of varying tones, intervals, and transpositions, layered on top of an amalgamation of harmonies, rhythms, dynamics, and timbres, all ensure a veritably infinite number of combinations and permutations. John Stuart Mill need not worry about the exhaustibility of music.

Music is exquisite and requisite. It is the creative juice that greases the machinery for insight and serendipity. It unlocks the right brain, opening a door to novel domains. It constructs neural networks and fosters pattern recognition. It trains willpower and discipline. Playing music builds a bigger and better highway between the two sides of the brain. Its myelinated interconnectivity nurtures out-of-the-box thinking and simulates meditation and flow. Like dreaming, it allows our powerful unconscious mind to sift and filter all the accumulated detritus into a meaningful story.

Sherlock Holmes and Albert Einstein epitomize these effects. But the characters are legion, and many others could have equally stood in for them: Plato and Pythagoras, Galileo and Kepler, Nietzsche and Schopenhauer, Planck and Heisenberg,

Helmholtz and Meitner. They all exemplify the vital contribution of music to creative thought. *Homo sapiens* evolved as ancient musicians, and we were all born into a musical heritage. Crucially, without music, *humanity* might not be so *sapient*.

BIBLIOGRAPHY

Abert, Hermann. *W. A. Mozart*. [1867].Translated by Stewart Spencer. New Haven: Yale University Press, 2007.

Adeney, Jennifer Marion. "European Starling (*Sturnus vulgaris*)." *Introduced Species Summary Project, Columbia University*, October 18, 2001. http://www.columbia.edu/itc/cerc/danoff-burg/invasion_bio/inv_spp_summ/Sturnus_vulgaris.html.

Agnihotri, Neeraj. *Procrasdemon: The Artist's Guide to Liberation from Procrastination*. Independently Published, 2019.

Altenmüller, Eckart, Wilfried Gruhn, Dietrich Parlitz, and Gundhild Liebert. "The Impact of Music Education on Brain Networks: Evidence from EEG-Studies." *International Journal of Music Education* 35, 1 (2000): 47–53. https://doi.org/10.1177/025576140003500115.

Anderson, Richard, ed. *The Pianist's Craft 2: Mastering the Works of More Great Composers*. Lanham, MD: Rowman & Littlefield Publishers, 2015.

Andrews, Evan. "Did a Premature Obituary Inspire the Nobel Prize?" *History*, July 23, 2020. https://www.history.com/news/did-a-premature-obituary-inspire-the-nobel-prize.

Avison, Charles. *An Essay on Musical Expression.* 3rd ed. London: Royal Society, 1775. Retrieved from https://quod.lib.umich.edu/e/ecco/004865013.0001.000?view=toc.

Azevedo, Frederico, Ludmila Carvalho, Lea Grinberg, José Marcelo Farfel, Renata Ferretti, Renata Leite, Wilson Jacob Filho, Roberto Lent, and Suzana Herculano-Houzel. "Equal Numbers of Neuronal and Nonneuronal Cells Make the Human Brain an Isometrically Scaled-Up Primate Brain." *The Journal of Comparative Neurology* 513, 5 (2009): 532–541.

Balbag, M. Alison, Nancy L. Pedersen, and Margaret Gatz. "Playing a Musical Instrument as a Protective Factor Against Dementia and Cognitive Impairment: A Population-Based Twin Study." *International Journal of Alzheimer's Disease* (2014): 836748. https://doi.org/10.1155/2014/836748.

Baptista, Luis, and Robin Keister. "Why Birdsong is Sometimes Like Music." *Perspectives in Biology & Medicine* 48, 3 (2005): 426-443. https://doi.org/10.1353/pbm.2005.0066.

Barnum, Barbara Stevens. "Why Freud and Jung Can't Speak: A Neurological Proposal." *Journal of Religion and Health* 45, 3 (2006): 346–58. http://www.jstor.org/stable/27512943.

Barrow, John D. *The World Within the World.* Oxford: Clarendon Press, 1988.

Barry, Rebecca. "Einstein's Violin Hit a High Note When It Sold for More Than a Half-Million Dollars at Auction," *Strings Magazine,* August 1, 2018. https://stringsmagazine.com/einsteins-violin-hit-a-high-note-when-it-sold-for-more-than-a-half-million-dollars-at-auction/.

Belkin, B. M., and F. A. Neelon. "The Art of Observation: William Osler and the Method of Zadig." *Annals of internal medicine* 116,10 (1992): 863-6. https://doi.org/10.7326/0003-4819-116-10-863

Bell, Joseph. "The Adventures of Sherlock Holmes: A Review." *The Bookman* 3 (1892): 79-81.

Bengtsson, Sara L., Zoltán Nagy, Stefan Skare, Lea Forsman, Hans Forssberg, and Fredrik Ullén. "Extensive Piano Practicing Has Regionally Specific Effects on White Matter Development." *Nature Neuroscience* 8 (2005): 1148-1150. https://doi.org/10.1038/nn1516.

Berlyne, D. E. *Aesthetics and Psychobiology.* New York: Appleby Century Crofts, 1971.

Blacking, John. *"A Commonsense View of All Music": Reflections on Percy Grainger's Contribution to Ethnomusicology and Music Education.* Cambridge, UK: Cambridge University Press, 1989.

Blake, William. *The Marriage of Heaven and Hell.* [1790]. Mineola, NY: Dover, 1984

Bloom, Allan. *The Closing of the American Mind.* New York: Simon & Schuster, 2008.

Bloom, Peter. *The Cambridge Companion to Berlioz.* Cambridge, UK: Cambridge University Press, 2000.

Boatman, D., J. Freeman, E. Vining, M. Pulsifer, D. Miglioretti, R. Minahan, B. Carson, J. Brandt, and G. McKhann. "Language Recovery After Left Hemispherectomy in Children with Late-Onset Seizures." *Annals of Neurology* 46 (1999): 579-586. https://doi.org/10.1002/1531-8249(199910)46:4%3C579::aid-ana5%3E3.0.co;2-k.

Boethius, Anicius Manlius Severinus. *Fundamentals of Music.* [ca. 1492]. Translated by Calvin M. Bower. New Haven: Yale University Press, 1989.

Bogen, J. E., and H. W. Gordon. "Musical Tests for Functional Lateralization with Intracarotid Amobarbital." *Nature* 230, 5295 (1971): 524-525. https://doi.org/10.1038/230524a0.

Bohm, David. *On Creativity*. Edited by Lee Nichol. London: Routledge Classics, 2003.

Borroff, Edith. "[Review of] *George Crumb: Profile of a Composer*." *American Music* 6, 1 (1988): 102–104. https://doi.org/10.2307/3448354.

Brackenridge, J Bruce. "Kepler, Elliptical Orbits, and Celestial Circularity: A Study in the Persistence of Metaphysical Commitment." *Annals of Science* 39, 3 (1982): 265-295. https://doi.org/10.1080/00033798200200241.

Breger, Louis. *Freud: Darkness in the Midst of Vision*. New York: Wiley & Sons, 2000.

Brooker, Jewel Spears, ed. *The Placing of T.S. Eliot*. Columbia, MO: University of Missouri Press, 1991.

Brooks, Michael. "There's a Glitch at the Edge of the Universe That Could Remake Physics." *New Scientist*, October 6, 2018.

Brothers, Harlan J. "Structural Scaling in Bach's Cello Suite No. 3." *Fractals* 15, 1 (2007): 89-95. https://doi.org/10.1142/S0218348X0700337X.

Brown, Brandon R. *Planck: Driven by Vision, Broken by War*. Oxford: Oxford University Press, 2015.

Browne, Malcolm W. "Left-Handed Universe." *New York Times*, November 25, 1986. https://www.nytimes.com/1986/11/25/science/left-handed-universe.html.

Brust, J. C. "Music and Language: Musical Alexia and Agraphia." *Brain* 103, 2 (1980): 367–392. https://doi.org/10.1093/brain/103.2.367.

Bucky, Peter, Albert Einstein, and Allen Weakland. *The Private Albert Einstein*. Kansas City, MO: Andrews and McMeel, 1992.

Bugos, J. A., W. M. Perlstein, C. S. McCrae, T. S. Brophy, and P. H. Bedenbaugh. "Individualized Piano Instruction Enhances Executive Functioning and Working Memory in Older Adults." *Aging and Mental Health* 11, 4 (2007): 464-471. https://doi.org/10.1080/13607860601086504.

Bullock, T. H., and R. Barrett. "Radiant Heat Reception in Snakes." *Communications in Behavioral Biology, part A*, 1 (1969): 19-29.

Burges, Virginia. "The Astonishing 300 Year History of the Gibson ex-Huberman Stradivarius." *Rhap.so.dy in Words*, August 24, 2015. www.rhapsodyinwords.com/2015/08/24/the-astonishing-300-year-history-of-the-gibson-ex-huberman-stradivarius/.

Burunat, Iballa, Elvira Brattico, Tuomas Puoliväli, Tapani Ristaniemi, Mikko Sams, and Petri Toiviainen. "Action in Perception: Prominent Visuo-Motor Functional Symmetry in Musicians during Music Listening." *PLoS ONE* 10, 9 (2015): e0138238. https://doi.org/10.1371/journal.pone.0138238.

Caen, Herb. "Hello, Out There." *San Francisco Chronicle*, July 3, 1996.

Cage, John. *Silence: 50th Anniversary Edition*. Middletown, CT: Wesleyan University Press, 2011.

Calaprice, Alice. *The Ultimate Quotable Einstein*. Princeton: Princeton University Press, 2011.

Carse, James P. *Finite and Infinite Games*. New York: Free Press, 1986.

Casey, Nell. "William Styron, 1925-2006: Unlikely Bard of Depression." *Slate*, November 7, 2006. https://www.slate.com/news-and-politics/2006/11/william-styron-unlikely-bard-of-depression.html.

Chatwin, Bruce. *The Songlines*. New York: Penguin Books, 1988.

Chomsky, Noam. *Syntactic Structures*. The Hague: Mouton, 1957.

Clarke-Ezzidio, Harry. "Venus Isn't Habitable—and It Could be All Jupiter's Fault." *CNN*, October 1, 2020. https://www.cnn.com/2020/10/01/world/venus-habitable-jupiter-scli-intl-scn/index.html.

Clark, Philip. "The Tuning Wars: 'Equal Temperament Destroys Everything...'" *Gramophone*, December 4, 2013. https://www.gramophone.co.uk/features/article/the-tuning-wars-equal-temperament-destroys-everything.

Clark, Ronald. *Einstein: The Life and Times*. New York: Harper Collins, 1971.

Cole, Rufus. "Dr. Osler: Scientist and Teacher." *Archives of Internal Medicine* 84, 1 (1949): 54–63. https://doi.org/10.1001/archinte.1949.00230010064013.

Colman, David R. "The Three Princes of Serendip: Notes on a Mysterious Phenomenon."*McGill Journal of Medicine* 9, 2 (2006): 161–163.

Conard, Nicholas, Maria Malina, and Susanne C. Münzel. "New Flutes Document the Earliest Musical Tradition in Southwestern Germany." *Nature* 460, (2009): 737-740. https://doi.org/10.1038/nature08169.

Conde-Valverde, Mercedes, Ignacio Martínez, Rolf M. Quam, Manuel Rosa, Alex D. Velez, Carlos Lorenzo, Pilar Jarabo, José María Bermúdez de Castro, Eudald Carbonell, and Juan Luis Arsuaga. "Neanderthals and *Homo sapiens* Had Similar Auditory and Speech Capacities." *Nature Ecology and Evolution* 5 (2021): 609–615. https://doi.org/10.1038/s41559-021-01391-6.

Coyle, Daniel. *The Talent Code*. New York: Bantam Books, 2009.

Cranston, Maurice. *Jean-Jacques - the Early Life and Work of Jean-Jacques Rousseau 1712-1754*. Chicago: University of Chicago Press, 1991.

Crelinsten, Jeffrey. *Einstein's Jury: The Race to Test Relativity.* Princeton: Princeton University Press, 2006.

Cross, Ian. "Music, Cognition, Culture, and Evolution." *Annals of the New York Academy of Sciences* 930 (2001): 28-42. https://doi.org/10.1111/j.1749-6632.2001.tb05723.x

Csikszentmihalyi, Mihaly. *Flow: The Psychology of Optimal Experience.* New York: Harper Perennial, 1991.

Dantzig, Tobias. *Number: The Language of Science.* New York: Plume, 2007.

Darwin, Charles. *The Expression of the Emotions in Man and Animals.* [1872]. Oxford: Oxford University Press, 2002.

Davis, Peter G. "A Virtuoso of Frightening Perfection." *New York Times,* June 29, 1975. https://www.nytimes.com/1975/06/29/archives/recordings-view-a-virtuoso-of-frightening-perfection.html.

Davoli, Silvia. "The Creation of the Word *Serendipity*." *Art, Law & More,* July 2, 2018. https://www.artlawandmore.com/2018/07/02/the-creation-of-the-word-serendipity/.

de Groot, Adriaan G. *Thought and Choice in Chess.* [1965]. The Hague: Mouton Publishers, 1978.

Delacroix, Eugène. *The Journal of Eugene Delacroix.* Edited and Translated by Walter Pach. New York: Hacker Art Books, 1980.

Delbert, Caroline. "A Supercomputer Just Calculated Pi to a Record-Breaking 62.8 Trillion Digits. So What?" *Popular Mechanics,* August 18, 2021. https://www.popularmechanics.com/science/math/a37329769/supercomputer-calculated-pi-to-record-breaking-628-trillion-digits/.

del Regato, Juan A. "Niels Bohr." *International Journal of Radiation Oncology Biology Physics* 7, 4 (1981): 509-529. https://doi.org/10.1016/0360-3016(81)90138-3.

Deutsch, Diana, Trevor Henthorn, Elizabeth Marvin, and HongShuai Xu. "Absolute Pitch Among American and Chinese Conservatory Students: Prevalence Differences, and Evidence for a Speech-Related Critical Period." *The Journal of the Acoustical Society of America* 119, 2 (2006): 719-722. https://doi.org/10.1121/1.2151799.

Diamond, Marian C., Arnold B. Scheibel, Greer M. Murphy Jr., and Thomas Harvey. "On the Brain of a Scientist: Albert Einstein." *Experimental Neurology* 88, 1 (1985): 198-204. https://doi.org/10.1016/0014-4886(85)90123-2.

Diedrich, Cajus G. 2015. "'Neanderthal Bone Flutes': Simply Products of Ice Age Spotted Hyena Scavenging Activities on Cave Bear Cubs in European Cave Bear Dens." *Royal Society Open Science* 2, 4 (2015): 140022. https://doi.org/10.1098/rsos.140022.

Dietrich, Arne. "Neurocognitive Mechanisms Underlying the Experience of Flow." *Consciousness and Cognition* 13, 4 (2004): 746-761. https://doi.org/10.1016/j.concog.2004.07.002.

Dijkgraaf, Robbert. "The World of Tomorrow." In *The Usefulness of Useless Knowledge,* by Abraham Flexner. Princeton: Princeton University Press, 2007.

Director, Bruce. "Riemann for Anti-Dummies," part 15. *LYM Canada*, February 20, 2010. http://lymcanada.org/15/.

Doyle, Arthur Conan. *The Adventures of Sherlock Holmes, and Other Stories.* San Diego: Canterbury Classics, 2011.

Doyle, Arthur Conan. *The Case-Book of Sherlock Holmes.* [1927]. Cornwall, UK: House of Stratus, 2008.

Doyle, Arthur Conan. *Memories and Adventures*. Cambridge, UK: Cambridge University Press, 1924.

Dubal, David, and Yehudi Menuhin. *Conversations with Menuhin: A Celebration on His 75th Birthday*. New York: Random House, 1991.

Duff, William. *An Essay on Original Genius*. London, 1767. https://ia600307.us.archive.org/33/items/essayonoriginalg00duff/essayonoriginalg00duff.pdf.

Dunnington, Waldo. "The Sesquicentennial of the Birth of Gauss." *The Scientific Monthly* 24, 5 (1927): 402–414. https://www.jstor.org/stable/7912.

du Sautoy, Marcus. "Listen by Numbers: Music and Maths." *Guardian*, June 27, 2011. https://www.theguardian.com/music/2011/jun/27/music-mathematics-fibonacci.

Dyson, Freeman. "Prof. Hermann Weyl, For.Mem.R.S." *Nature* 177 (1956): 457–458. https://doi.org/10.1038/177457a0.

Dyson, Freeman. *Disturbing the Universe*. New York: Harper and Row, 1979.

Eckert, Michael. *Arnold Sommerfeld: Science, Life and Turbulent Times 1868-1951*. New York: Springer Publishing, 2013.

Edelman, Gerald. *Bright Air, Brilliant Fire: On the Matter of the Mind*. New York: Basic Books, 1991.

Einstein, Albert. *Albert Einstein Archives*. [1923-1955]. Hebrew University of Jerusalem. http://www.albert-einstein.org/.index.html.

Einstein, Albert. *Ideas and Opinions*. New York: Crown Publishers, 1954.

Einstein, Albert. *Autobiographical Notes*. Edited by Paul Schilpp. Chicago: Open Court Publications, 1979.

Elbert, Thomas, Christo Pantev, Christian Wienbruch, Brigitte Rockstroh, and Edward Taub. "Increased Cortical Representation of the Fingers of the Left Hand in String Players." *Science* 270, 5234 (1995): 305-307. https://doi.org/10.1126/science.270.5234.305.

Ernst, Sabine. *Lise Meitner an Otto Hahn: Briefe aus den Jahren 1912 bis 1924*. Stuttgart: Wissenschaftliche Verlagsgesellschaft, 1992.

Etinson, Adam. "Is a Life without Struggle Worth Living?" *The New York Times*, October 2, 2017. https://www.nytimes.com/2017/10/02/opinion/js-mill-happiness-anxiety.html.

Fatović-Ferenčić, Stella, and Karl Holubar. ""...They Were Always Making Discoveries, by Accidents and Sagacity, of Things They Were Not in Quest of..." Serendipity: a Quarter of a Millennium Anniversary." *Journal of Investigative Dermatology* 121, 4 (2003): xiii-xiv. https://doi.org/10.1046/j.1523-1747.2003.12505.x.

Fehér, Olga, Haibin Wang, Sigal Saar, Partha P. Mitra, and Ofer Tchernichovski. "De Novo Establishment of Wild-Type Song Culture in the Zebra Finch." *Nature* 459 (2009): 564-568. https://doi.org/10.1038/nature07994.

Feynman, Richard. *The Feynman Lectures on Physics: New Millennium Edition*. [1963]. New York: Basic Books, 2011.

Feynman, Richard. *"What Do You Care What Other People Think?"* New York: W. W. Norton & Company, 1988.

Fiore, Thomas M. "Music and Mathematics." June 3, 2004. https://www-personal.umd.umich.edu/~tmfiore/1/musictotal.pdf.

Flexner, Abraham. *The Usefulness of Useless Knowledge*. [originally in *Harper's Magazine*, 1939]. Princeton: Princeton University Press, 2017.

Foldes, Lili. "A Musical Visit with Einstein." *The Etude Magazine* 65, 1 (1947).

Folger, Tim. "Why Quantum Mechanics Still Stumps Physicists." *Discover*, April 11, 2017. https://www.discovermagazine.com/the-sciences/why-quantum-mechanics-still-stumps-physicists.

Forgeard, Marie, Gottfried Schlaug, Andrea Norton, Camilla Rosam, Udita Iyengar, and Ellen Winner. "The Relation Between Music and Phonological Processing in Normal-Reading Children and Children with Dyslexia." *Music Perception: An Interdisciplinary Journal* 25, 4 (2008): 383–90. https://doi.org/10.1525/mp.2008.25.4.383.

Foster, Brian. "CERN, the violin and the music of the spheres." *CERN Courier,* February 1, 2005. https://www.cerncourier.com/a/cern-the-violin-and-the-music-of-the-spheres/.

Foucault, Michel. "The Masked Philosopher," in *Foucault Live: Interviews, 1966–1984,* 2nd ed. Edited by Sylvère Lotringer. Translated by Lysa Hochroth. New York: Semiotext(e), 1996.

Fox, Matthew. *A Spirituality Named Compassion*. Rochester, VT: Inner Traditions, 1999.

Frank, Philipp. *Einstein: His Life and Times*. Translated by George Rosen. New York: Knopf, 1963.

Freud, Harry. "My Uncle Sigmund." In *Freud as We Knew Him*. Edited by Hendrik Ruitenbeek. Detroit: Wayne State University Press, 1973.

Freud, Sigmund. "The Economic Problem of Masochism." [1924]. In *The Standard Edition of the Complete Psychological Works of Sigmund Freud*. Translated by James Strachey. London: Hogarth Press, 1961.

Frisch, Otto. *What Little I Remember*. [1979]. Cambridge, UK: Cambridge University Press, 1991.

Fujioka, Takako, Bernhard Ross, Ryusuke Kakigi, Christo Pantev, and Laurel J. Trainor. "One Year of Musical Training Affects Development of Auditory-Cortical-Evoked Fields in Young Children." *Brain* 129, 10 (2006): 2593-2608. https://doi.org/10.1093/brain/awl247.

Galilei, Galileo. *Dialogues Concerning Two New Sciences.* Translated by Henry Crew and Alfonso de Salvio. New York: Dover, 1914.

Gay, Peter. *Freud: A Life of Our Time.* New York: W.W. Norton, 2006.

Gilbertson, Mark, Martha E. Shenton, Aleksandra Ciszewski, Kiyoto Kasai, Natasha B. Lasko, Scott P. Orr, and Roger K. Pitman. "Smaller Hippocampal Volume Predicts Pathologic Vulnerability to Psychological Trauma." *Nature Neuroscience* 5, 11 (2002): 1242–1247. https://doi.org/10.1038/nn958.

Gingras, Bruno. "Johannes Kepler's *Harmonices Mundi*: A "Scientific" Version of the Harmony of the Spheres." *Journal of the Royal Astronomical Society of Canada* 97, 6 (2003): 259-265. http://adsabs.harvard.edu/pdf/2003JRASC..97..259G.

Gladwell, Malcolm. *Outliers: The Story of Success.* New York: Little, Brown and Company, 2008.

Gleick, James. *Genius: The Life and Science of Richard Feynman.* New York: Vintage Books, 1992.

Gombrich, Ernst. *The Sense of Order: A Study in the Psychology of Decorative Art.* New York: Phaidon Press, 1984.

Grant, Adam. *Think Again: The Power of Knowing What You Don't Know.* New York: Viking, 2021.

Gray, Jeremy. *Henri Poincaré: A Scientific Biography.* Princeton: Princeton University Press, 2013.

Grinstein, Louise, and Sally Lipsey. *Encyclopedia of Mathematics Education.* Oxfordshire: Taylor & Francis, 2001.

Hadamard, Jacques. *The Mathematician's Mind: The Psychology of Invention in the Mathematical Field.* Princeton: Princeton University Press, 2020.

Haldane, J. B. S. *Possible Worlds and Other Essays*. London: Chatto and Windum, 1927.

Halperin, Tamar. "The Ongoing Quest for Bach's Temperament." *Juilliard Journal,* March 2009. http://journal.juilliard.edu/journal/ongoing-quest-bachs-temperament.

Hambrick, David, Frederick Oswald, Erik Altmann, Elizabeth Meinz, Fernand Gobet, Guillermo Campitelli. "Deliberate Practice: Is That All It Takes to Become an Expert?" *Intelligence* 45 (2014): 34-45. https://doi.org/10.1016/j.intell.2013.04.001.

Harari, Yuval Noah. *Sapiens: A Brief History of Humankind*. New York: Harper Collins, 2014.

Hardy, G H. *A Mathematician's Apology*. Cambridge, UK: Cambridge University Press, 1940.

Hassin, Ran, James S. Uleman, and John A. Bargh, eds. *The New Unconscious*. New York: Oxford University Press, 2005.

Hayes, John R. "Three Problems in Teaching General Skills." In *Thinking and Learning Skills*. Vol. 2. Edited by Susan F. Chipman, Judith W. Segal, and Robert Glaser. New York: Routledge, 2013.

Heilbron, J. L. *The Dilemmas of an Upright Man: Max Planck and the Fortunes of German Science*. Cambridge, MA: Harvard University Press, 2000.

Henahan, Donal. "Did Shostakovich Have a Secret?" *New York Times*, July 10, 1983. https://www.nytimes.com/1983/07/10/arts/music-view-did-shostakovich-have-a-secret.html.

Heraclitus. *Fragments: The Collected Wisdom of Heraclitus*. [ca. 535-475 BCE]. Translated by Brooks Haxton. New York: Penguin Books, 2003.

Herrnstein, Richard, and Charles Murray. *The Bell Curve: Intelligence and Class Structure in American Life.* New York: Free Press, 2010.

Hill, Peter. *Stravinsky: The Rite of Spring.* Cambridge, UK: Cambridge University Press, 2000.

Hoeft, Fumiko, Bruce D. McCandliss, Jessica M. Black, Alexander Gantman, Nahal Zakerani, Charles Hulme, Heikki Lyytinen, et al. "Neural Systems Predicting Long-Term Outcome in Dyslexia." *Proceedings of the National Academy of Sciences* 108, 1 (2011): 361–366. https://doi.org/10.1073/pnas.1008950108.

Hofstadter, Douglas R. *Gödel, Escher, Bach: An Eternal Golden Braid.* New York: Basic Books, 1979.

Hofstadter, Douglas R. *Le Ton Beau de Marot: In Praise of the Music of Language.* New York: Basic Books, 1997.

Hölzel, Britta K., James Carmody, Mark Vangel, Christina Congleton, Sita M. Yerramsetti, Tim Gard, and Sara W. Lazar. "Mindfulness practice leads to increases in regional brain gray matter density." *Psychiatry Research* 191, 1 (2011): 36–43. https://dx.doi.org/10.1016%2Fj.pscychresns.2010.08.006.

Horgan, John. "Physics Titan Still Thinks String Theory Is "On the Right Track."" *Scientific American*, September 22, 2014. https://blogs.scientificamerican.com/cross-check/physics-titan-still-thinks-string-theory-is-on-the-right-track/.

Horgan, John. "What Is Philosophy's Point? Part 5—A Call for "Negative Philosophy."" *Scientific American*, February 10, 2017. https://blogs.scientificamerican.com/cross-check/what-is-philosophys-point-part-5-a-call-for-negative-philosophy/

Horgan, John. "Is the Schrödinger Equation True?" *Scientific American*, January 7, 2021. https://www.scientificamerican.com/article/is-the-schroedinger-equation-true1/

Horowitz, Alexandra. *On Looking*. New York: Scribner, 2013.

Horridge, Adrian. "How Bees Distinguish Colors." *Eye Brain* 7 (2015): 17-34. https://dx.doi.org/10.2147%2FEB.S77973.

Huxley, Aldous. *The Perennial Philosophy*. New York: Harper & Brothers, 1945.

Huxley, Thomas. "On the Method of Zadig." *Popular Science Monthly* 17 (1880): 467-475. Retrieved from https://www.gutenberg.org/cache/epub/2627/pg2627-images.html.

Hyde, Krista, Robert J. Zatorre, Timothy D. Griffiths, Jason P. Lerch, and Isabelle Peretz. "Morphometry of the Amusic Brain." *Brain* 129, 10 (2006): 2562-2570. https://doi.org/10.1093/brain/awl204.

Iliescu, Delia Monica Duca. "The Impact of Artificial Intelligence on the Chess World." *JMIR Serious Games* 8,4 (2020): e24049. https://doi.org/10.2196/24049.

Isaacson, Walter. *Einstein: His Life and Universe*. New York: Simon & Schuster, 2007.

Jacobsen, Jörn-Henrik, Johannes Stelzer, Thomas Hans Fritz, Gael Chételat, Renaud La Joie, and Robert Turner. "Why Musical Memory Can Be Preserved in Advanced Alzheimer's Disease." *Brain* 138, 8 (2015): 2438–2450. https://doi.org/10.1093/brain/awv135.

JAMA Editors. "The Method of Zadig in Medicine." *Journal of the American Medical Association* 18, 4 (1892): 111-113. https://doi.org/10.1001/jama.1892.02411080021008.

James, William. *The Energies of Men*. New York: Moffat, Yard and Company, 1914.

Jensen, Claudia R. "A Theoretical Work of Late Seventeenth-Century Muscovy: Nikolai Diletskii's *Grammatika* and the Earliest Circle of

Fifths." *Journal of the American Musicological Society* 45, 2 (1992): 305–331. https://doi.org/10.2307/831450.

Johnson, Craig. "Functions of Number Theory in Music." *Mathematics Teacher* 94, 8 (2001): 700-707.

Jordania, Joseph. *Why Do People Sing? Music in Human Evolution*. West Beach, South Australia: Logos, 2011.

Jorgensen-Earp, Cheryl, and Darwin Jorgensen. "Miracle from Mouldy Cheese: Chronological versus Thematic Self-Narratives in the Discovery of Penicillin." *Quarterly Journal of Speech* 88, 1 (2002): 69–90. https://doi.org/10.1080/00335630209384360.

Jourdain, Robert. *Music, the Brain, and Ecstasy: How Music Captures Our Imagination*. New York: W. Morrow, 1997.

Jung-Beeman, Mark, Edward M. Bowden, Jason Haberman, Jennifer L Frymiare, Stella Arambel-Liu, Richard Greenblatt, Paul J Reber, and John Kounios. "Neural Activity When People Solve Verbal Problems with Insight." *PLoS Biology* 2, 4 (2004): E97. https://doi.org/10.1371/journal.pbio.0020097.

Jung, Carl. *Memories, Dreams, Reflections*. [1961]. New York: Vintage Books, 1989.

Kac, Mark. *Enigmas of Chance: An Autobiography*. New York: Harper Collins, 1985.

Kafka, Franz. *Letters to Friends, Family and Editors*. Translated by Richard Winston and Clara Winston. New York: Schocken Books, 1977.

Kahneman, Daniel. *Thinking, Fast and Slow*. New York: Farrar, Straus, and Giroux, 2011.

Kalmijn, A. J. "The Electric Sense of Sharks and Rays." *Journal of Experimental Biology* 55 (1971): 371-383. https://doi.org/10.1242/jeb.55.2.371.

Kaminsky, Jack. *Hegel on Art: An Interpretation of Hegel's Aesthetics.* New York: State University of New York Press, 1962.

Kant, Immanuel. *Critique of Judgment.* Edited and Translated by Paul Guyer. Cambridge, UK: Cambridge University Press, 2000.

Keats, John. "On a Grecian Urn." In *Annals of Fine Art.* Vol. 4. [1820]. *Hathitrust,* Retrieved November 27, 2021. https://babel.hathitrust.org/cgi/pt?id=mdp.39015012982370&view=1up&seq=7&skin=2021.

Kepler, Johannes. *The Harmony of the World.* [1619]. Translated by Alistair Matheson Duncan, E. J. Aiton, and Judith Veronica Field. Philadelphia: American Philosophical Society, 1997.

Key, J. D., and A. E. Rodin. "Medical Reputation and Literary Creation: An Essay on Arthur Conan Doyle versus Sherlock Holmes 1887-1987." *Adler Museum Bulletin* 13, 2 (1987): 21-25.

Kim, Kyung Hee. "The Creativity Crisis: The Decrease in Creative Thinking Scores on the Torrance Tests of Creative Thinking." *Creativity Research Journal* 23, 4 (2011): 285-295. https://doi.org/10.1080/10400419.2011.627805.

Knight, Ruth, Mizanur Khondoker, Nicholas Magill, Robert Stewart, and Sabine Landau. "A Systematic Review and Meta-Analysis of the Effectiveness of Acetylcholinesterase Inhibitors and Memantine in Treating the Cognitive Symptoms of Dementia." *Dementia and Geriatric Cognitive Disorders* 45, 3-4 (2018): 131-151. https://doi.org/10.1159/000486546.

Ko, Yoree. "Stradivarius Nets $16M for Japan Quake Relief". *The Wall Street Journal,* June 21, 2011. https://blogs.wsj.com/scene/2011/06/21/stradivarius-nets-16m-for-japan-quake-relief/.

Kobbé, Gustav. "Wagner, Minna and Cosima." *The North American Review* 174, 543 (1902): 259-268. https://www.jstor.org/stable/25105292.

Koelsch, Stefan, Thomas Fritz, Katrin Schulze, David Alsop, and Gottfried Schlaug. "Adults and Children Processing Music: An fMRI Study." *Neuroimage* 25, 4 (2005): 1068-1076. https://doi.org/10.1016/j.neuroimage.2004.12.050.

Koelsch, Stefan. "Brain Correlates of Music-Evoked Emotions." *Nature Reviews of Neuroscience* 15, 3 (2014): 170–180. https://doi.org/10.1038/nrn3666.

Koestler, Arthur. *The Act of Creation*. New York: The Macmillan Company, 1964.

Konnikova, Maria. *Mastermind: How to Think Like Sherlock Holmes*. New York: Penguin Books, 2013.

Kotler, Steven. *The Rise of Superman: Decoding the Science of Ultimate Human Performance*. Boston: New Harvest/Houghton Mifflin Harcourt, 2014.

Kotler, Steven. *The Art of Impossible: A Peak Performance Primer*. (Kindle Edition). New York: Harper Wave Books, 2021.

Kounios, John, and Mark Beeman. *The Eureka Factor: Aha Moments, Creative Insights, and the Brain*. New York: Random House, 2015.

Kuhn, Thomas S. *The Essential Tension: Selected Studies in Scientific Tradition and Change*. Chicago: University of Chicago Press, 2011.

Laënnec, René. *A Treatise on the Diseases of the Chest and on Mediate Auscultation*. Translated by Sir John Forbes. Philadelphia: Samuel Wood & Sons, 1830. Retrieved from https://collections.nlm.nih.gov/ext/mhl/9308215/PDF/9308215.pdf.

Langer, Ellen. "The Illusion of Control." *Journal of Personality and Social Psychology* 32, 2 (1975): 311–328. https://doi.org/10.1037/0022-3514.32.2.311.

Langer, Susanne. *Philosophy in a New Key*, 3rd ed. Cambridge, MA: Harvard University Press, 1957.

Lau, Beverly, Bonnie Blackwell, Henry Schwarcz, Ivan Turk, and Joel Blickstein. "Dating a Flautist? Using ESR (Electron Spin Resonance) in the Mousterian Cave Deposits at Divje Babe I, Slovenia." *Geoarchaeology-an International Journal* 12, 6 (1997): 507-536. https://doi.org/10.1002/(SICI)1520-6548(199709)12:6%3C507::AID-GEA2%3E3.0.CO;2-2.

Laylin, James, ed. *Nobel Laureates in Chemistry, 1901-1992*. United States: American Chemical Society, 1993.

Lazar, Sara, Catherine E. Kerr, Rachel H. Wasserman, Jeremy R. Gray, Douglas N. Greve, Michael T. Treadway, Metta McGarvey, et al. "Meditation Experience Is Associated with Increased Cortical Thickness." *Neuroreport* 16, 17 (2005): 1893–1897. https://doi.org/10.1097/01.wnr.0000186598.66243.19.

Lebovic, Nitzan. "Dionysian politics and the discourse of "Rausch."" *UCLA Working Papers*, 2004. Retrieved from https://escholarship.org/uc/item/9z91f2vs.

Le Guin, Ursula. *Very Far Away from Anywhere Else*. Orlando, FL: Harcourt, 2004.

Lehman, Bradley. "Bach's Extraordinary Temperament: Our Rosetta Stone." *Early Music* 33, 1 (2005): 3-23. https://muse.jhu.edu/article/181841.

Lendvai, Ernö. *Béla Bartók: An Analysis of His Music*. London: Kahn & Averill Publishers, 1971.

Levitin, Daniel. *This is Your Brain on Music: The Science of a Human Obsession*. New York: Plume, 2006.

Lewis, Wilmarth S. "The Yale Edition of Horace Walpole's Correspondence, 1933-1973." *The Yale University Library Gazette* 48, 2 (1973): 69-83. http://www.jstor.org/stable/40858508.

Liébert, Georges. *Nietzsche and Music*. Chicago: University of Chicago Press, 2004.

Lightman, Alan. *The Accidental Universe*. New York: Vintage Books, 2013.

Lightman, Alan. *In Praise of Wasting Time*. New York: Simon & Schuster, TED books, 2018.

Lippard, Lucy R. *Overlay: Contemporary Art and the Art of Prehistory*. New York: Pantheon Books, 1983.

Loewi, Otto. "Über Humorale Übertragbarkeit der Herznervenwirkung." *Pflügers Archiv für die Gesamte Physiologie des Menschen und der Tiere* 204 (1924): 629-640. https://doi.org/10.1007/BF01738910.

Lokhorst, Gert-Jan. "Hemisphere differences before 1800." *Behavioral and Brain Sciences* 8, 4 (1985): 642-642. https://doi.org/10.1017/S0140525X0004543X.

Lokhorst, Gert-Jan. "The First Theory about Hemispheric Specialization: Fresh Light on an Old Codex." *Journal of the History of Medicine and Allied Sciences* 51, 3 (1996): 293-312. https://doi.org/10.1093/jhmas/51.3.293.

Low, Barbara. Psycho-Analysis: A Brief Account of the Freudian Theory. London: Allen & Unwin, 1920.

Löwenstein, Hubertus. *Towards the Further Shore*. London: Victor Gollancz, 1968.

Loy, Gareth. *Musimathics: The Mathematical Foundations of Music*. Cambridge, MA: MIT Press, 2011.

Madrigal, Alexis. "What Was Steve Jobs's High School GPA? Not 4.0, or even 3.0." *The Atlantic Magazine*, February 9, 2012. https://www.theatlantic.com/technology/archive/2012/02/what-was-steve-jobss-high-school-gpa-not-40-or-even-30/252828/.

Mandelbrot, Benoit. *The Fractal Geometry of Nature*. New York: Macmillan, 1982.

Marden, Orison Swett. *How They Succeeded*. Boston: Lothrop Publishing Company, 1901.

Marinkovic, Ksenija, Sharelle Baldwin, Maureen G. Courtney, Thomas Witzel, Anders M. Dale, and Eric Halgren. "Right Hemisphere Has the Last Laugh: Neural Dynamics of Joke Appreciation." *Cognitive, Affective & Behavioral Neuroscience* 11, 1 (2011): 113–130. https://dx.doi.org/10.3758%2Fs13415-010-0017-7.

Market Folly. "Notes from Charlie Munger's Daily Journal Meeting 2015." *Market Folly*, March 26, 2015. https://www.marketfolly.com/2015/03/notes-from-charlie-mungers-daily.html.

Matsui, Toshie, Satomi Tanaka, Koji Kazai, Minoru Tsuzaki, and Haruhiro Katayose. "Activation of the left superior temporal gyrus of musicians by music-derived sounds." *Neuroreport* 24, 1 (2013): 41-45. https://doi.org/10.1097/wnr.0b013e32835c1e02.

Maurer, Charles., and Daphne Maurer. *The World of the Newborn*. New York: Basic Books, 1988.

McCoy, Alli N., and Yong Siang Tan. "Otto Loewi (1873-1961): Dreamer and Nobel laureate." *Singapore medical journal* 55,1 (2014): 3-4. https://doi.org/10.11622/smedj.2014002.

McGuire, William, and Carl Jung. *C. G. Jung Speaking: Interviews and Encounters*. Princeton: Princeton University Press, 1987.

McLellan, Joseph. "Charles Rosen, Dissecting Beethoven Sonatas." *Washington Post*, June 28, 1992. https://www.washingtonpost.

com/archive/lifestyle/style/1992/06/28/charles-rosen-dissecting-beethoven-sonatas/9353647c-63fc-4c0b-bce0-578990114c6e/.

Menuhin, Yehudi. *Theme and Variations*. New York: Stein and Day, 1972.

Milius, Susan. "Music without Borders: When Birds Trill and Whales Woo-oo, We Call It Singing. Are We Serious?" *Science News* 157, 16 (2000): 252-254. https://doi.org/10.2307/4012494.

Mill, John Stuart. *A System of Logic, Ratiocinative and Inductive, Being a Connected View of the Principles of Evidence, and the Methods of Scientific Investigation*. [1851]. Frankfurt: Outlook Verlag, 2020.

Mill, John Stuart. *Autobiography*. 2nd ed. London: Longmans, Green, Reader, and Dyer, 1873.

Miller, Geoffrey. "Evolution of human music through sexual selection." In *The Origins of Music*. Edited by Nils Lennart Wallin, Björn Merker, and Steven Brown, 329-360. Cambridge, MA: MIT Press, 2000.

Mithen, Steven. *The Singing Neanderthals: The Origins of Music, Language, Mind, and Body*. Cambridge, MA: Harvard University Press, 2006.

MIT Sloan Sports Analytics Conference. "SSAC14: 10,000 Hours vs. The Sports Gene." *YouTube*, March 10, 2014. Video, 1:03:15. https://www.youtube.com/watch?v=iXBhINPoKEk.

Monk, Ray. *Ludwig Wittgenstein: The Duty of Genius*. New York: The Free Press, 1990.

Moore, Samuel. "Superaccurate GPS Chips Coming to Smartphones in 2018." *IEEE Spectrum: Technology, Engineering, and Science News*, September 21, 2017. https://spectrum.ieee.org/tech-talk/semiconductors/design/superaccurate-gps-chips-coming-to-smartphones-in-2018.

Moreno, Sylvain, Ellen Bialystok, Raluca Barac, E. Glenn Schellenberg, Nicholas J. Cepeda, and Tom Chau. "Short-Term Music Training Enhances Verbal Intelligence and Executive Function." *Psychological Science* 22, 11 (2011): 1425–1433. https://dx.doi.org/10.1177%2F0956797611416999.

Morgan, W. W. "A Morphological Life." *Annual Reviews of Astronomy and Astrophysics* 26 (1988): 1-9.

Morris, Edmund. *Edison*. New York: Random House, 2019.

Nabokov, Vladimir. *Speak, Memory: An Autobiography Revisited*. [1951]. New York: Vintage Books, 1989.

Nagel, Jody. "The Pythagorean Comma." *JOMAR Press*, March 13, 2003. https://www.jomarpress.com/nagel/articles/PythagoreanComma.html.

Nantais, Kristin, and E. Glenn Schellenberg. "The Mozart Effect: An Artifact of Preference." *Psychological Science* 10, 4 (1999): 370–373. https://doi.org/10.1111%2F1467-9280.00170.

Nestler, Eric J. "The Neurobiology of Cocaine Addiction." *Science & Practice Perspectives* 3, 1 (2005): 4-10. https://doi.org/10.1151/spp05314.

Neville, Morgan, dir. *Won't You Be My Neighbor?* Los Angeles: Tremolo Productions, 2018. Video, 93 minutes.

New England Journal of Medicine. "Cycling for Freezing Gait in Parkinson's Disease." *YouTube*, April 1, 2010. Video, 1:49. https://www.youtube.com/watch?v=aaY3gz5tJSk.

Newlands, John. *The Discovery of the Periodic Law*. London: E & F. N. Spon, 1884.

Newman, Ernest. *A Musical Motley*. London: John Lane, 1919.

New York Times. "Relative Tide and Sand Bars Trap Einstein; He Runs His Sailboat Aground at Old Lyme." *New York Times*, August 4, 1935. https://www.nytimes.com/1935/08/04/archives/relative-tide-and-sand-bars-trap-einstein-he-runs-his-sailboat.html.

New York Times. "Why Did Heifetz Fluff? "Just Forgot," He Admits." *New York Times*, January 11, 1954. https://www.nytimes.com/1954/01/11/archives/why-did-heifetz-fluff-just-forgot-he-admits.html.

Nietzsche, Friedrich. *Twilight of the Idols*. [1889]. Oxford: Oxford University Press, 2008.

Nietzsche, Friedrich. *The Will to Power*. [1901]. Overland Park, KS: Digireads.com, 2010.

Nobel Media. "The Symphony of Science." *NobelPrize.org*, January 23, 2020. https://www.nobelprize.org/symphony-of-science/.

Norton, A. A. "Proficiency in Shorthand Writing." *Western Stenographer* 1, 7 (1894): 13. https://babel.hathitrust.org/cgi/pt?id=nyp.33433034364822&view=1up&seq=127.

Nottebohm, Fernando. "Testosterone Triggers Growth of Brain Vocal Control Nuclei in Adult Female Canaries." *Brain Research* 189, 2 (1980): 429-436.

Nova. "A Radical Mind." *Nova*, September 30, 2008. https://www.pbs.org/wgbh/nova/article/mandelbrot-fractal/.

Ohno, Susumu, and Midori Ohno. "The All Pervasive Principle of Repetitious Recurrence Governs Not Only Coding Sequence Construction but Also Human Endeavor in Musical Composition." *Immunogenetics* 24, 2 (1986): 71-78. https://doi.org/10.1007/BF00373112.

Oldfield, Molly. *The Secret Museum: Some Treasures Are Too Precious to Display*. Ontario: Firefly Books, 2013.

Oppenheimer, J. Robert. *Robert Oppenheimer: Letters and Recollections.* [1935]. Edited by Alice Smith and Charles Weiner. Cambridge, MA: Harvard University Press, 1980.

Orwell, George. *Nineteen Eighty-Four.* Oxford: Oxford University Press, 2021.

Osler, William. "The Natural Method of Teaching the Subject of Medicine." *Journal of the American Medical Association* 36, 24 (1901): 1673-1679. https://doi.org/10.1001/jama.1901.52470240001001.

Osler, William. "On the Educational Value of the Medical Society." *Boston Medical and Surgical Journal* 148, 11 (1903): 275-279. https://doi.org/10.1056/NEJM190303121481101.

Overy, Katie. "Dyslexia, Temporal Processing and Music: The Potential of Music as an Early Learning Aid for Dyslexic Children." *Psychology of Music* 28, 2 (2000): 218-229. https://doi.org/10.1177%2F0305735600282010.

Page, Tim. "Unpretentious Prodigy Puzzled by All the Fuss." *New York Times*, July 29, 1986. https://www.nytimes.com/1986/07/29/arts/unpretentious-prodigy-puzzled-by-all-the-fuss.html.

Pais, Abraham. *Subtle is the Lord.* Oxford: Oxford University Press, 2005.

Pammer, Kristen. "Brain Mechanisms and Reading Remediation: More Questions than Answers." *Scientifica* 2014 (2014): 802741. https://doi.org/10.1155/2014/802741.

Parker, Ian. "Yuval Noah Harari's History of Everyone, Ever." *New Yorker*, February 10, 2020. https://www.newyorker.com/magazine/2020/02/17/yuval-noah-harari-gives-the-really-big-picture.

Patel, Sutchin, Sara Best, and Ronald Rabinowitz. "Sherlock Holmes and the Case of the Vanishing Examination." *American Journal of Medicine* 131, 11 (2018): 1270-1271. https://doi.org/10.1016/j.amjmed.2018.05.015.

Paterniti, Michael. *Driving Mr. Albert.* New York: Random House, 2013.

Peschel, R. E., and E. Peschel. "What Physicians Have in Common with Sherlock Holmes." *Journal of the Royal Society of Medicine* 82, 1 (1989): 33-36. https://doi.org/10.1177/014107688908200114.

Pesic, Peter. *Music and the Making of Modern Science.* Cambridge, MA: MIT Press, 2014.

Phillpotts, Eden. *A Shadow Passes.* London: Cecil Palmer & Hayward, 1918.

Piccirilli, Massimo, Tiziana Sciarma, and Simona Luzzi. 2000. "Modularity of Music: Evidence from a Case of Pure Amusia." *Journal of Neurology, Neurosurgery & Psychiatry* 69, 4 (2000): 541-545. https://dx.doi.org/10.1136%2Fjnnp.69.4.541.

Planck, Max. *Where is Science Going?* Translated by James Murphy. New York: W.W. Norton, 1932.

Plato. *Plato: The Collected Dialogues.* [ca. 428-348 BCE]. Edited by Edith Hamilton and Huntington Cairns. Translated by Lane Cooper, et al. Princeton: Princeton University Press, 1961.

Podlech, Joachim. ""Try and Fall Sick …"—The Composer, Chemist, and Surgeon Aleksandr Borodin." *Angewandte Chemie International Edition* 49, 37 (2010): 6490-6495. https://doi.org/10.1002/anie.201002023.

Poe, Edgar Allan. "The Murders in the Rue Morgue." In *The Complete Tales & Poems of Edgar Allan Poe.* Edison, NJ: Castle Books, 2002.

Poe, Edgar Allan. *The Mystery of Marie Rogêt.* New York: R. F. Fenno & Company, 1899.

Poincaré, Henri. *The Foundations of Science: Science and Hypothesis, The Value of Science, Science and* Method. Edited by J. McKeen Cattell. Translated by George Bruce Halsted. New York: Science Press, 1913.

Popper, Karl. *Conjectures and Refutations.* London: Routledge, 2002.

Powers, Richard. *Orfeo: A Novel.* New York: W. W. Norton & Company, 2014.

Pressfield, Steven. *The War of Art.* Brooklyn: Black Irish Entertainment, 2012.

Raglio, Alfredo, Osmano Oasi, Marta Gianotti, Agnese Rossi, Karine Goulene, and Marco Stramba-Badiale. "Improvement of Spontaneous Language in Stroke Patients with Chronic Aphasia Treated with Music Therapy: A Randomized Controlled Trial." *International Journal of Neuroscience* 126, 3 (2016): 235-242. https://doi.org/10.3109/00207454.2015.1010647.

Raglio, Alfredo, Alberto Zaliani, Paola Baiardi, Daniela Bossi, Cinzia Sguazzin, Edda Capodaglio, Chiara Imbriani, Giulia Gontero, and Marcello Imbriani. "Active music therapy approach for stroke patients in the post-acute rehabilitation." *Neurological Sciences* 38, 5 (2017): 893–897.

Ramón y Cajal, Santiago. *Degeneration & Regeneration of the Nervous System.* Vol. 2. [1928]. Edited and Translated by Raoul May. New York: Hafner Publishing Company, 1959.

Rauscher, Frances, Gordon Shaw, and Catherine Ky. "Music and Spatial Task Performance." *Nature* 365 (1993): 611. https://doi.org/10.1038/365611a0.

Reed, James. "A Medical Perspective on the Adventures of Sherlock Holmes." *Journal of Medical Ethics* 27 (2001): 76-81. http://dx.doi.org/10.1136/mh.27.2.76.

Reston, James, Jr. *Galileo: A Life*. Washington, DC: Beard Books, 2000.

Reverberi, Carlo, Alessio Toraldo, Serena D'Agostini, and Miran Skrap. "Better without (Lateral) Frontal Cortex? Insight Problems Solved by Frontal Patients." *Brain* 128, 12 (2005): 2882-2890. https://doi.org/10.1093/brain/awh577.

Richards, Stan, David Culp, and Archie Richards, Jr. *The Peaceable Kingdom: Building a Company Without Factionalism, Fiefdoms, Fear and Other Staples of Modern Business*. New York: John Wiley & Sons, 2001.

Richardson, W. John, Charles R. Greene, Jr., Charles I. Malme, and Denis H. Thomson. *Marine Mammals and Noise*. Edited by W. John Richardson. San Diego: Academic Press, 1995.

Rizzi, Sofia. "What Is the Fibonacci Sequence - and Why Is it the Secret to Musical Greatness?" *Classic FM*, January 28, 2019. https://www.classicfm.com/discover-music/fibonacci-sequence-in-music/.

Roberts, Gareth E. *From Music to Mathematics: Exploring the Connections*. Baltimore: Johns Hopkins University Press, 2016.

Rogenmoser, Lars, Julius Kernbach, Gottfried Schlaug, and Christian Gaser. "Keeping Brains Young with Making Music." *Brain Structure and Function* 223, 1 (2018): 297–305. https://doi.org/10.1007/s00429-017-1491-2.

Root-Bernstein, Robert, Maurine Bernstein, and Helen Garnier. "Correlations between Avocations, Scientific Style, Work Habits, and Professional Impact of Scientists." *Creativity Research Journal* 8, 2 (1995): 115-137. https://doi.org/10.1207/s15326934crj0802_2.

Root-Bernstein, Robert. "Music, Creativity and Scientific Thinking." *Leonardo* 34, 1 (2001): 63-68. https://doi.org/10.1162/002409401300052532.

Rosado, P. "The Invisible Colours of the Universe: Gravitational Waves—Fast Forward Science 2013." *YouTube*, August 31, 2013. Video 4:59. https://www.youtube.com/watch?v=SVHm3tAjnNQ.

Rosen, Charles. *The Classical Style: Haydn, Mozart, Beethoven*. New York: W. W. Norton, 1972.

Rothman, Lily. "Here's What Beethoven Did When He Lost His Hearing." *Time Magazine*, December 17, 2015. https://www.time.com/4152023/beethoven-birthday/.

Rothstein, Edward. "Math and Music: The Deeper Links. *New York Times*, August 29, 1982. https://www.nytimes.com/1982/08/29/arts/math-and-music-the-deeper-links.html.

Rothstein, Edward. *Emblems of Mind: The Inner Life of Music and Mathematics*. Chicago: University of Chicago Press, 2006.

Rovelli, Carlo. *Seven Brief Lessons on Physics*. New York: Riverhead Books, 2014.

Sacks, Oliver. *Musicophilia: Tales of Music and the Brain*. New York: Alfred A. Knopf, 2008.

Sarlo, Dario. *The Performance Style of Jascha Heifetz*. London: Routledge, 2016.

Schellenberg, E. Glenn. "Music Lessons Enhance IQ." *Psychological Science* 15, 8 (2004): 511–514. https://doi.org/10.1111%2Fj.0956-7976.2004.00711.x.

Scherer, John. "Before Cardiac MRI: René Laënnec and the Invention of the Stethoscope." *Cardiology Journal* 14, 5 (2007): 518-519.

Schlaug, Gottfried, Lutz Jäncke, Yanxiong Huang, Jochen F. Staiger, and Helmuth Steinmetz. "Increased Corpus Callosum Size in Musicians." *Neuropsychologia* 33, 8 (1995): 1047-1055. https://doi.org/10.1016/0028-3932(95)00045-5.

Schopenhauer, Arthur. *Parerga and Paralipomena: Short Philosophical Essays*. [1851]. Translated by E. F. J. Payne. Oxford: Clarendon Press, 2000.

Schopenhauer, Arthur. *The World as Will and Representation*. Vol. 1. [1819]. Translated by E. F. J. Payne. New York: Dover Publications, 1969.

Schrobsdorff, Susanna. "Teen Depression and Anxiety: Why the Kids Are Not Alright." *Time Magazine*, October 27, 2016. https://www.time.com/magazine/us/4547305/november-7th-2016-vol-188-no-19-u-s/.

Sedbrook, Danielle. "Must the Molecules of Life Always Be Left-Handed or Right-Handed?" *Smithsonian Magazine*, July 28, 2016. https://www.smithsonianmag.com/space/must-all-molecules-life-be-left-handed-or-right-handed-180959956/.

Seelig, Carl. *Albert Einstein: A Documentary Biography*. London: Staples Press, 1956.

Service, Tom. "Friedrich Nietzsche's Horrible Music." *BBC radio*, August 30, 2013. Audio, 20 minutes. https://www.bbc.co.uk/programmes/b038yjv0.

Shulman, Laurie. "Beethoven's *Pastoral*: Program Notes." *Jacksonville Symphony*, 2019. https://www.jaxsymphony.org/beethoven-pastoral-program-notes/.

Sime, Ruth Lewin. *Lise Meitner: A Life in Physics*. Berkeley: University of California Press, 1996.

Singh, Parmanand. "The So-Called Fibonacci Numbers in Ancient and Medieval India." *Historia Mathematica* 12 (1985): 229-244.

Snowdon, David A. "Aging and Alzheimer's Disease: Lessons from the Nun Study." *The Gerontologist* 37, 2 (1997): 150-156. https://doi.org/10.1093/geront/37.2.150.

Sommerfeld, Arnold. *Atomic Structure and Spectral Lines*. Translated by Henry Leopold Brose. New York: E. P. Dutton and Company, 1923.

Sontag, Susan. *Styles of Radical Will*. New York: Picador, 2002.

Sperber, Dan, and Ann Sperber. *Explaining Culture: A Naturalistic Approach*. Oxford: Blackwell Publishers, 1996.

Sperry, Roger. "Roger W. Sperry—Nobel Lecture: Some Effects of Disconnecting the Cerebral Hemispheres." *NobelPrize.org*, December 8, 1981. https://www.nobelprize.org/prizes/medicine/1981/sperry/25059-roger-w-sperry-nobel-lecture-1981/.

Sperry, Roger. "Some Effects of Disconnecting the Cerebral Hemispheres." *Science* 217, 4566 (1982): 1223-1226. https://doi.org/10.1126/science.7112125.

Stachel, John, Diana Kormos Buchwald, David Cassidy, Robert Schulmann, Jürgen Renn, Martin Klein, A. J. Knox, et al., eds. *The Collected Papers of Albert Einstein*. Volumes 1-15. [1879-1922] Princeton: Princeton University Press, 1987-2017.

Stainer, Cecie. *A Dictionary of Violin Makers*. London: Novello, 1896. Retrieved from https://www.ebooksread.com/authors-eng/c-cecie-stainer/a-dictionary-of-violin-makers-iat.shtml.

Storr, Anthony. *Music and the Mind*. New York: Ballantine Books, 1993.

Strathern, Paul. *Mendeleyev's Dream: The Quest for the Elements.* New York: St. Martin's Press, 2001.

Stravinsky, Igor, and Robert Craft. *Conversations with Igor Stravinsky.* New York: Doubleday & Company, 1959.

Stravinsky, Igor, and Robert Craft. *Dialogues.* Berkeley: University of California Press, 1982.

Styron, William. *Darkness Visible: A Memoir of Madness.* New York: Random House, 1990.

Südhof, Thomas C. "Thomas C. Südhof: Biographical." *NobelPrize.org,* January 23, 2020. https://www.nobelprize.org/prizes/medicine/2013/sudhof/biographical/.

Sullivan, J.W.N. "The Justification of the Scientific Method." *Athenaeum,* May 1919.

Sutherland, John. "Should They De-Nobel Moniz?" *Guardian,* August 2, 2004. https://www.theguardian.com/education/2004/aug/02/highereducation.comment.

Suzuki, D. T. *Zen Buddhism: Selected Writings of D. T. Suzuki.* Edited by William Barrett. New York: Three Leaves Press, 2006.

Suzuki, Shinichi. *Nurtured by Love: A New Approach to Education.* [1969]. Translated by Waltraud Suzuki. USA: Alfred Publishing Co., 1983.

Taleb, Nassim. *Antifragile: Things That Gain from Disorder.* New York: Random House, 2012.

Taylor, Jill Bolte. *My Stroke of Insight: A Brain Scientist's Personal Journey.* New York: Viking, 2008.

Taylor, Wendell. "J. A. R. Newlands: A Pioneer in Atomic Numbers." *Journal of Chemical Education* 26, 9 (1949): 491-496. https://doi.org/10.1021/ed026p491.

Thoreau, Henry David. *Walden, or Life in the Woods.* Boston: Houghton, Mifflin, 1892.

Tu, Chau. "Seeing the Patterns of Sound." *Science Friday*, November 7, 2016. https://www.sciencefriday.com/articles/seeing-the-patterns-in-sound/.

Understanding Evolution Team. "Early Evolution and Development: Ernst Haeckel, Evolution 101." *University of California Museum of Paleontology*, Retrieved February 16, 2020. https://evolution.berkeley.edu/evolibrary/article/0_0_0/history_15.

van 't Hoff, Jacobus Henricus. *Imagination in Science.* [1878]. Edited by Arnost Kleinzeller and Georg F. Springer. Translated by Georg F. Springer. Berlin: Springer-Verlag, 1967.

Van Wagenen, W. P., and R. Y. Herren. "Surgical Division of Commissural Pathways in the Corpus Callosum: Relation to Spread of an Epileptic Attack." *Archives of Neurology & Psychiatry* 44 (1940): 740–759. https://doi.org/10.1159/000098832.

Vermetten, Eric, Meena Vythilingam, Steven M. Southwick, Dennis S. Charney, and J. Douglas Bremner. "Long-Term Treatment with Paroxetine Increases Verbal Declarative Memory and Hippocampal Volume in Posttraumatic Stress Disorder." *Biological Psychiatry* 54, 7 (2003): 693–702. https://doi.org/10.1016/s0006-3223(03)00634-6.

Viereck, George Sylvester. "What Life Means to Einstein." *Saturday Evening Post*, October 26, 1929. https://www.saturdayeveningpost.com/wp-content/uploads/satevepost/what_life_means_to_einstein.pdf.

Vitruvius. *The Ten Books on Architecture*. [ca. 70-15 BCE]. Translated by Morris Hicky Morgan. Cambridge, MA: Harvard University Press, 1914. https://gutenberg.org/cache/epub/20239/pg20239-images.html.

Vollman, Henning, Patrick Ragert, Virginia Conde, Arno Villringer, Joseph Classen, Otto W. Witte, and Christopher J. Steele. "Instrument Specific Use-Dependent Plasticity Shapes the Anatomical Properties of the Corpus Callosum: A Comparison Between Musicians and Non-Musicians." *Frontiers in Behavioral Neuroscience* 8 (2014): 245. https://doi.org/10.3389/fnbeh.2014.00245.

Wainwright, Milton, and John Wainwright. *Miracle Cure: The Story of Penicillin and the Golden Age of Antibiotics*. Oxford: Blackwell Publishers, 1990.

Waldrop, Mitch. "Inside Einstein's Love Affair with "Lina"—His Cherished Violin." *National Geographic Magazine*, February 3, 2017. https://www.nationalgeographic.com/news/2017/02/einstein-genius-violin-music-physics-science/.

Wallas, Graham. *The Art of Thought*. London: Jonathan Cape, 1926.

Ward, Jamie. "Synesthesia." *Annual Review of Psychology* 64, 1 (2013): 49-75. https://doi.org/10.1146/annurev-psych-113011-143840.

Wasserstein, A. "Theaetetus and the History of the Theory of Numbers." *The Classical Quarterly* 8, 3-4 (1958): 165-179. https://doi.org/10.1017/S0009838800021790.

Westmoreland, Barbara, and J. D. Key. "Arthur Conan Doyle, Joseph Bell, and Sherlock Holmes. A Neurologic Connection." *Archives of Neurology* 48 3 (1991): 325-329. https://doi.org/10.1001/ARCHNEUR.1991.00530150097025.

Weyl, Hermann. *Philosophy of Mathematics and Natural Sciences*. Princeton: Princeton University Press, 1949.

Whitehead, Alfred North. *Science and the Modern World.* Cambridge, UK: Cambridge University Press, 1925.

Whitrow, G. J. *Einstein, the Man and His Achievement: A Series of Broadcast Talks Under the General Editorship of G. J. Whitrow.* London: BBC, 1967.

Wilson, Edward O. *The Meaning of Human Existence.* New York: W.W. Norton & Company, 2014.

Wolf, Gary. "Steve Jobs: The Next Insanely Great Thing." *Wired Magazine,* February 1, 1996. https://www.wired.com/1996/02/jobs-2/.

Wolff, Christoph. *The New Bach Reader.* [1945]. Edited by Hans David and Arthur Mendel. New York: W.W. Norton & Company, 1999.

Yanes, Javier. "Euler, the Beethoven of Mathematics." *OpenMind,* April 9, 2018. https://www.bbvaopenmind.com/en/science/leading-figures/euler-the-beethoven-of-mathematics/.

Young, John. *What the Robin Knows.* New York: Houghton Mifflin Harcourt, 2012.

Zegers, Richard H. C. "The Eyes of Johann Sebastian Bach." *Archives of Ophthalmology* 123, 10 (2005): 1427–1430. https://doi.org/10.1001/archopht.123.10.1427.

Zuckerkandl, Victor. *Sound and Symbol: Music and the External World.* Vol. 1. Translated by Willard R. Trask. Princeton: Princeton University Press, 1969.

IMAGE ATTRIBUTIONS

The cover image and the images used at the beginning of each chapter are adapted from Gerd Altmann. Pixabay, pixabay.com/illustrations/annoy-cells-stars-dendrites-sepia-4051659/

Figure 1. Solvay Conference, 1927. Pixabay, pixabay.com/photos/einstein-physicist-conference-67711/.

Figure 2. Albert Einstein, 1921. Harris and Ewing Studio. National Portrait Gallery, Smithsonian Institution; gift of Aileen Conkey.

Figure 3. Circle of fifths. By Douglas Wadle.

Figure 4. Albert Einstein, Adolf Hurwitz, und seine Tochter Lisbeth Hurwitz, 1913. *ETH-Bibliothek Zürich, Bildarchiv* / Fotograf: Unbekannt / Portr_07389 / Public Domain Mark, doi.org/10.3932/ethz-a-000046051.

Figure 5. Pablo de Sarasate. Die Gartenlaube. 1886. Issue No. 7. Germany: Ernst Keil.

Figure 6. Einstein, Albert (1879–1955), Königin Elisabeth von Belgien (1876–1965) in Laeken (Belgien). 1932. ETH-Bibliothek Zürich, Bildarchiv / Fotograf: Albert I, König von Belgien / Portr_03076 / Public Domain Mark, ba.e-pics.ethz.ch/#1618337313744_4.

Figure 7. Bain News Service, P. (ca. 1915) Tuscha [sic] Seidel., ca. 1915. Library of Congress, www.loc.gov/item/2014706397/.

Figure 8. Donald Glaser on viola. Circa 1960s–1970. Courtesy of The Caltech Archives. Donald Glaser Papers, Box 1.9, Folder 6.

Figure 9. Thomas Edison. [Between 1870 and 1880]. Library of Congress, www.loc.gov/item/2017893349/.

Figure 10. Image by Michael Gaida. Pixabay, pixabay.com/photos/piano-keys-music-piano-keyboard-3138964/.

Figure 11. Image by Pexels. Pixabay, pixabay.com/photos/music-violin-musical-instrument-1283851/.

Figure 12. Image by Gerd Altmann. Pixabay, pixabay.com/illustrations/neurons-brain-cells-brain-structure-582054/.

Figure 13. Image by Nicky Hayes. Pixabay, pixabay.com/vectors/motor-neurone-neuron-neurone-nerve-2040692/.

Figure 14. Image by Vlad Vasnetsov. Pixabay, pixabay.com/photos/piano-key-inside-piano-keys-music-3957653/.

Figure 15. Bain News Service, P. 1917. Jascha Heifetz. Library of Congress, www.loc.gov/item/2014705777/.

Figure 16. Image by Pexels. Pixabay, pixabay.com/photos/antique-violins-classical-music-1868064/.

Figure 17. Yehudi Menuhin, 1932. Portrait by Mark S. Joffe. National Portrait Gallery, Smithsonian Institution; gift of *Time* magazine, music.si.edu/object/npg_NPG.88.TC141.

Figure 18. Image by Ri Butov. Pixabay, pixabay.com/photos/cello-tuning-key-classic-musical-4412332/.

Figure 19. Einstein, Albert (1879–1955) als Segler an der Pinne. 1934. *ETH-Bibliothek Zürich, Bildarchiv* / Fotograf: Unbekannt / Portr_03124 / Public Domain Mark, doi.org/10.3932/ethz-a-000045762.

Figure 20. Leonard Bernstein, half-length portrait, facing right, seated at piano, making annotations to musical score. 1955. *World Telegram & Sun* photo by Al Ravenna. Library of Congress, www.loc.gov/item/2001695796/.

Figure 21. John Cage, 1988. Netherlands National Archives, CC∅, www.nationaalarchief.nl/onderzoeken/fotocollectie/ad6b69c0-d0b4-102d-bcf8-003048976d84.

Figure 22. Drawing by the author.

Figure 23. Drawing by the author.

Figure 24. Motor homunculus. Betts, J. 2013. *Anatomy and Physiology*. Open Stax. Houston, TX, openstax.org/books/anatomy-and-physiology/pages/1-introduction.

Figure 25. Chiral Molecules may have Hitched Rides to Planets. August 20, 2018. NASA, astrobiology.nasa.gov/news/chiral-molecules-may-have-hitched-rides-to-planets/.

Figure 26. Drawing by the author.

Figure 27. Image by Ferrand. Pixabay, pixabay.com/photos/fern-american-hart-s-tongue-fern-439595/.

Figure 28. Image by Vizetelly. Pixabay, pixabay.com/illustrations/nightingale-bird-songbird-5880255/.

Figure 29. Beddard, F. E. 1903. On the modifications of structure in the syrinx of the Accipitres, with remarks on other points in anatomy of that group. *Proceedings of the Zoological Society of London*.

157–163, archive.org/details/proceedingsofzoo19032zool/page/157/mode/1up?view=theater.

Figure 30. Mandelbrot Fractal. Image by Nepomuk-si. Pixabay, pixabay.com/illustrations/fractal-fractals-background-pattern-2839218/.

Figure 31. Adapted from sunflower image by Casey Pilley. Pixabay, pixabay.com/photos/flower-sunflower-plant-disk-flowers-94187/.

Figure 32. Image by Nicolas Damian Visceglio. Pixabay, pixabay.com/vectors/fibonacci-spiral-science-golden-1601158/.

Figure 33. Johann Sebastian Bach. ~1750. Library of Congress, www.loc.gov/item/2004671945/.

Figure 34. Bach, J. S. circa 1742-1749. The Art of Fugue, BWV 1080. Berlin State Library, resolver.staatsbibliothek-berlin.de/SBB0001D9A000000000.

Figure 35. Harris & Ewing, photographer. Albert Einstein with wife Elsa; State, War, and Navy building in background. Washington, DC, United States. [Between 1921 and 1923]. Library of Congress, www.loc.gov/item/2016885960/.

Figure 36. Chladni, Ernst Florens Friedrich. 1787. Entdeckungen uber die Theorie des Klanges mit eilf Kupfertafeln., Leipzig : Bey Weidmanns Erben und Reich, library.si.edu/image-gallery/99550.

Figure 37. Image by Brent Connelly. Pixabay, pixabay.com/photos/wood-instrument-vintage-wooden-3130608/.

Figure 38. Bach, J. S. 1722–1723. *The Well-Tempered Clavier*, title page. Berlin State Library, resolver.staatsbibliothek-berlin.de/SBB000189D400000000.

Figure 39. Image by Public Domain Archive. Pixabay, pixabay.com/photos/piano-grand-piano-musical-instrument-349928/.

NOTES

AEA: *Albert Einstein Archives.* (Einstein, A.) Listed with the relevant folder, document.

TCPAE : *The Collected Papers of Albert Einstein.* (Stachel, et al.) Listed with the relevant volume, document.

Overture: Accidental Sagacity

1. Altenmüller, "Impact of Music Education," 51.
2. Colman, "Three Princes of Serendip,"161.
3. Jorgensen-Earp, "Miracle from Mouldy Cheese," 78.
4. Wainwright, *Miracle Cure,* 102.
5. Flexner, *Usefulness of Useless Knowledge,* 57
6. Foucault, "Masked Philosopher," 305.
7. Davoli, "Creation of the Word *Serendipity.*"
8. Lewis, *Horace Walpole's Correspondence,* 407.
9. Lewis, *Horace Walpole's Correspondence,* 408.
10. Fatović-Ferenčić, "Serendipity," xiii.
11. Belkin, "Art of Observation," 863.
12. JAMA, "Method of Zadig in Medicine," 113.
13. Huxley, "On the Method of Zadig," 467-475.
14. Belkin, "Art of Observation," 865.
15. Osler, "Natural Method of Teaching," 1673-74.
16. Osler, "Educational Value of the Medical Society," 325.
17. Cole, "Dr. Osler," 60.
18. Patel, "Sherlock Holmes," 1270.
19. Bell, "Adventures of Sherlock Holmes," 79-81.

[20] Bell, "Adventures of Sherlock Holmes," 79-81.
[21] Westmoreland, "Arthur Conan Doyle," 326.
[22] Doyle, *Memories and Adventures*, 25.
[23] Bell, "Adventures of Sherlock Holmes," 79-81.
[24] Key, "Medical Reputation," 21-25.
[25] Reed, "A Medical Perspective," 77.
[26] Poe, *Marie Rogêt*, 83.
[27] Poe, "Murders," in *Complete Tales*, 126-127.
[28] Doyle, "The Redheaded League," in *Adventures of Sherlock Holmes*, 181.
[29] Doyle, "A Study in Scarlet," in *Adventures of Sherlock Holmes*, 13.
[30] Bell, "Adventures of Sherlock Holmes," 79-81.
[31] Doyle, "The Boscombe Mystery," in *Adventures of Sherlock Holmes*, 213.
[32] Doyle, "The Reigate Puzzle," in *Adventures of Sherlock Holmes*, 417.
[33] Horowitz, *On Looking*, 11.
[34] Doyle, "The Hound of the Baskervilles," in *Adventures of Sherlock Holmes*, 712.
[35] Doyle, "The Blanched Soldier," in *Case-Book of Sherlock Holmes*, 36.
[36] Schopenhauer, *Parerga and Paralipomena*, volume 2, 110.
[37] Doyle, "Silver Blaze," in *Adventures of Sherlock Holmes*, 351.
[38] Viereck, "What Life Means to Einstein," in Calaprice, *Quotable Einstein*, 12.
[39] Einstein to Carl Seelig, (*AEA*, 1952), 39, 13.
[40] Einstein to Conrad and Paul Habicht, (*TCPAE*, 1907), 5, 56.
[41] Rovelli, *Seven Brief Lessons*, 36.
[42] Reed, "A Medical Perspective," 80.
[43] Peschel, "What Physicians Have in Common," 33.
[44] Bloom, *Companion to Berlioz*, 29.
[45] Podlech, "Try and Fall Sick," 6490.
[46] Root-Bernstein, "Music, Creativity and Scientific Thinking," 63-68.
[47] Laënnec, *Treatise on the Diseases of the Chest*, 5.
[48] Scherer, "Before Cardiac MRI," 518-519.
[49] Laënnec, *Treatise on the Diseases of the Chest*, 5.
[50] Laënnec, *Treatise on the Diseases of the Chest*, 66.
[51] Laënnec, *Treatise on the Diseases of the Chest*, (translator's note), 517.

[52] McLellan, "Dissecting Beethoven Sonatas."
[53] Rosen, *The Classical Style*, 76.
[54] Einstein, "Fundamental Ideas," (Unpublished article intended for *Nature* magazine). (*TCPAE*, 1920), 7, 31.
[55] Suzuki, *Nurtured by Love*, 78.
[56] Isaacson, *Einstein*, 262.
[57] Frank, *Einstein*, 101.
[58] Dijkgraaf, "World of Tomorrow," 19.
[59] Moore, "Superaccurate GPS Chips."
[60] Jensen, "Nikolai Diletskii's *Grammatika*," 305–331.

C: Introductions

[1] Isaacson, *Einstein*, 37.
[2] Isaacson, *Einstein*, 38.
[3] Einstein Response to Emil Hilb Questionnaire, (*AEA*, 1939), 34, 322.
[4] Foldes, "Musical Visit with Einstein," in Calaprice, *Quotable Einstein*, 239.
[5] Einstein to Siegfried Ochs, (*TCPAE*, 1924), 14, 373.
[6] Bucky, *Private Albert Einstein*, in Calaprice, *Quotable Einstein*, 239.
[7] Dijkgraaf, "World of Tomorrow," 17.
[8] Isaacson, *Einstein*, 7.
[9] Whitrow, "The Man and His Achievement," 21.
[10] Isaacson, *Einstein*, 14.
[11] Hofstadter, *Godel, Escher, Bach*, 175.
[12] Doyle, "A Study in Scarlet," in *Adventures of Sherlock Holmes*, 13.
[13] Mill, *System of Logic*, 153.
[14] Doyle, "A Study in Scarlet," in *Adventures of Sherlock Holmes*, 11.
[15] Doyle, "The Red-Headed League,"in *Adventures of Sherlock Holmes*, 183.
[16] Doyle, "The Cardboard Box," in *Adventures of Sherlock Holmes*, 940.
[17] Doyle, "The Red-Headed League," in *Adventures of Sherlock Holmes*, 181-182.
[18] Doyle, "The Red-Headed League," in *Adventures of Sherlock Holmes*, 183.
[19] Newman, *Musical Motley*, 263.
[20] Newman, *Musical Motley*, 277.

21. Newman, *Musical Motley*, 277.
22. Doyle, "A Study in Scarlet," in *Adventures of Sherlock Holmes*, 25.
23. Newman, *Musical Motley*, 277.
24. Doyle, "A Study in Scarlet," in *Adventures of Sherlock Holmes*, 16.
25. Doyle, "The Cardboard Box," in *Adventures of Sherlock Holmes*, 940.
26. Stainer, *Dictionary of Violin Makers*, 88.
27. Ko, "Stradivarius Nets $16M."
28. Burges, "Gibson ex-Huberman Stradivarius."
29. Foster, "Violin and the music of the spheres."
30. Barry, "Einstein's Violin Hit a High Note."
31. Doyle, "A Study in Scarlet," in *The Adventures of Sherlock Holmes*, 10.
32. Doyle, "A Study in Scarlet," in *The Adventures of Sherlock Holmes*, 30.

G: Laureates

1. Ryffe, "Inspector's report on a music examination," (*TCPAE*, 1896) 1, 17.
2. Isaacson, *Einstein*, 35.
3. Morris, *Edison*, 619.
4. Madrigal, "Steve Jobs's High School GPA."
5. Waldrop, "Einstein's Love Affair with Lina."
6. Isaacson, Einstein, 415.
7. Einstein to Queen Elisabeth. (*AEA*, 1936), 32, 387.
8. Isaacson, *Einstein*, 272.
9. Waldrop, "Einstein's Love Affair with Lina."
10. Isaacson, *Einstein*, 427.
11. Andrews, "Premature Obituary."
12. Dijkgraaf, "World of Tomorrow," 27.
13. Laylin, *Nobel Laureates*, 1.
14. van 't Hoff, *Imagination in Science*, 12-15.
15. Robert Oppenheimer to Frank Oppenheimer. *Letters and Recollections*, 190.
16. Root-Bernstein, "Correlations between Avocations," 115-137.
17. Root-Bernstein, "Music, Creativity and Scientific Thinking," 63-68.
18. Courant to Einstein, (*TCPAE*, 1922), 13, 19.

19. Isaacson, *Einstein*, 167.
20. Einstein to Elsa Einstein, (*TCPAE*, 1920), 10, 9.
21. Brown, *Planck*, 73.
22. Brown, *Planck*, 7.
23. Sime, *Lise Meitner*, 62.
24. Brown, *Planck*, 21-23.
25. Sime, *Lise Meitner*, 35.
26. Frisch, *What Little I Remember*, 33, 49.
27. Südhof, "Biographical."
28. Südhof, "Biographical."
29. Nobel Media. "Symphony of Science."
30. Nobel Media, "Symphony of Science."
31. Reston, *Galileo*, 7.
32. Sacks, *Musicophilia*, 240.
33. Brown, *Planck*, 58.
34. Morris, *Edison*, 155.
35. Rothman, "Here's What Beethoven Did."
36. Morris, *Edison*, 156.
37. Low, *Psycho-Analysis*, 73.
38. Freud, *Economic Problem of Masochism*, 159.
39. Gay, *Freud*, 316.
40. Breger, *Darkness in the Midst of Vision*, 32.
41. H. Freud, "My Uncle Sigmund," 313.

D: Quadrivium

1. Plato, "Republic," in *Collected Dialogues*, 646.
2. Plato, "Republic," in *Collected Dialogues*, 656.
3. Bloom, *Closing of the American Mind*, 71.
4. Schopenhauer, *World as Will and Representation*, 256.
5. Schopenhauer [quoting Leibniz, *Letters*, 154],*World as Will and Representation*, 256.
6. Schopenhauer, *World as Will and Representation*, 264.
7. Schopenhauer, *World as Will and Representation*, 257.
8. Schopenhauer, *World as Will and Representation*, 264.
9. Storr, *Music and the Mind*, 156.
10. Service, "Nietzsche's Horrible Music."
11. Service, "Nietzsche's Horrible Music."

[12] Liébert, *Nietzsche and Music*, 18.
[13] Liébert, *Nietzsche and Music*, 13.
[14] Nietzsche, *Twilight of the Idols*, 9.
[15] Sacks, *Musicophilia*, 258.
[16] Sacks, *Musicophilia*, 338.
[17] Rauscher, "Music and Spatial Task Performance," 611.
[18] Nantais, "Mozart Effect," 370–373.
[19] Moreno, "Short-term music training," 1425–1433.
[20] Schellenberg, "Music Lessons Enhance IQ," 511–514.
[21] Fujioka, "One Year of Musical Training," 2593-2608.
[22] Bugos, "Individualized Piano Instruction," 464-471.
[23] Ramón y Cajal, *Degeneration and Regeneration*, 750.
[24] Balbag, "Playing a Musical Instrument."
[25] Knight, "Effectiveness of Acetylcholinesterase Inhibitors," 135.
[26] Snowdon, "Aging and Alzheimer's Disease," 155.
[27] Rogenmoser, "Keeping Brains Young," 297–305.
[28] Jacobsen, "Musical Memory Can be Preserved, 2438–2450.
[29] Raglio, "Active Music Therapy," 893–897.
[30] Bugos, "Individualized Piano Instruction," 464-471.
[31] Market Folly, "Notes."
[32] Agnihotri, *Procrasdemon*, 24.
[33] Gladwell, *Outliers*, 35.
[34] MIT Sloan Sports, "10,000 Hours vs. *The Sports Gene*."
[35] Hambrick, "Deliberate Practice," 43.
[36] Hayes, "Three Problems," 396.
[37] Hadamard, *Mathematician's Mind*, 142.
[38] Wolf, "Steve Jobs."
[39] Koestler, *Act of Creation*, 35.
[40] Paterniti, *Driving Mr. Albert*, 47 & 190.
[41] Paterniti, *Driving Mr. Albert*, 51.
[42] Diamond, "Brain of a Scientist," 198-204.
[43] Paterniti, *Driving Mr. Albert*, 51.
[44] Azevedo, "Equal Numbers," 535.
[45] Levitin, *Your Brain on Music*, 88.
[46] Edelman, *Bright Air, Brilliant Fire*, 17.
[47] Coyle, *Talent Code*, 33, 61.
[48] Bengtsson, "Extensive Piano Practicing," 1148-1150.
[49] Coyle, *Talent Code*, 50-52.

50 Taleb, *Antifragile*, 3.
51 Norton, Shorthand Writing," 13.

A: Descent with Modification

1. Lokhorst, "Fresh Light on an Old Codex," 293.
2. Sacks, *Musicophilia*, 155.
3. Understanding Evolution, "Ernst Haeckel."
4. Wilson, *Meaning of Human Existence*, 9.
5. Boethius, *Fundamentals of Music*, Book 1, 8.
6. Sacks, *Musicophilia*, 347.
7. Lau, "Dating a Flautist?" 507-536.
8. Diedrich, "Neanderthal bone flutes," 1.
9. Conard, "New Flutes," 737-740.
10. Blacking, *Commonsense View*, 22.
11. Cranston, *Jean-Jacques Rousseau*, 289.
12. Darwin, *Expression of the Emotions in Man*, 92.
13. Doyle, "A Study in Scarlet," in *Adventures of Sherlock Holmes*, 26.
14. Conde-Valverde, "Neanderthals," 612.
15. Mithen, *Singing Neanderthals*, 26.
16. Harari, *Sapiens*, chapter 2, 22-44.
17. Jordania, *Why Do People Sing?* Chapter 1.
18. Deutsch, "Absolute Pitch," 719.
19. Chomsky, *Syntactic Structures*, xiv.
20. Sacks, *Musicophilia*, 152.
21. Chatwin, *Songlines*, 108.
22. de Groot, *Thought and Choice in Chess*, iv.
23. Davis, "Virtuoso of Frightening Perfection."
24. Sarlo, *Style of Jascha Heifetz*, 36.
25. New York Times, "Why Did Heifetz Fluff?"

E: Epiphany

1. Eckert, *Science, Life and Turbulent Times*, 41, 45, & 267.
2. Sommerfeld, *Atomic Structure*, v.
3. del Regato, "Niels Bohr," 517.
4. Duff, *Original Genius*, 9.
5. Duff, *Original Genius*, 6-7.
6. Kahneman, *Thinking, Fast and Slow*, 20-21.

7   Blake, *Marriage of Heaven and Hell*, Plate 10.
8   Bohm, *On Creativity*, 3.
9   Einstein to Erwin Freundlich, (*TCPAE*, 1930), 8, 123.
10 Einstein to Carl Seelig, (*AEA*, 1952), 39, 13; in Calaprice, *Quoting Einstein*, 20.
11 Dijkgraaf, "World of Tomorrow," 29.
12 Lightman, *Wasting Time*, 47.
13 Grant, *Think Again*, 24.
14 Gleick, *Genius*, 130.
15 Gleick, *Genius*, 320.
16 Kac, *Enigmas of Chance*, xxv.
17 Rothstein, *Emblems of Mind*, 192.
18 Rothstein, *Emblems of Mind*, 192.
19 Gleick, *Genius*, 285-286.
20 Gleick, *Genius*, 65.
21 Gleick, *Genius*, 15.
22 Gleick, *Genius*, 105.
23 Plato. "Timaeus," in *Collected Dialogues*, 1175.
24 Gleick, *Genius*, 16.
25 Gleick, *Genius*, 105.
26 Rovelli, *Seven Brief Lessons*, 44.
27 Kaminsky, *Hegel on Art*, 125.
28 Crelinsten, *Einstein's Jury*, 116.
29 Einstein, "Kyoto Lecture," (*TCPAE*, 1922), 13, 399.
30 Caen, "Hello," C1.
31 Bucky, *Private Albert Einstein*, 150.
32 Seelig, *Einstein*, 15.
33 Cage, *Silence*, 12.
34 Fox, *Spirituality Named Compassion*, 129.
35 Pressfield, *War of Art*, 110-115.
36 Whitrow, "The Man and His Achievement."
37 Vitruvius. *Ten Books on Architecture*, Book IX, 254, par. 10.
38 Cage, *Silence*, 72.
39 Morgan, "Morphological Life," 8.
40 Jung-Beeman, "Neural Activity," E97.
41 Kounios, *Eureka Factor*, 84-85.
42 Piccirilli, "Modularity of Music," 541-545.
43 Koelsch, "Processing Music," 1068-1076.

[44] Matsui, "Activation," 41-45.
[45] Kounios, *Eureka Factor*, 90.
[46] Wallas, *Art of Thought*, 80-82.
[47] Feynman, *Lectures on Physics*, Vol. 2, 20-10.
[48] Le Guin, *Very Far Away*, 44.

B: Disorder and Distraction

[1] Dubal, *Conversations with Menuhin*, 113.
[2] Menuhin, *Theme and Variations*, 9.
[3] Plato. "Timaeus," in *Collected Dialogues*, 1175.
[4] Koestler, *Act of Creation*, 352.
[5] Dyson, *Disturbing the Universe*, 8.
[6] Einstein to Heinrich Zangger, (*TCPAE*, 1914), 5, 513.
[7] Berlyne, *Aesthetics and Psychobiology*, 226.
[8] Rovelli, *Seven Brief Lessons*, 19.
[9] Folger, "Quantum Mechanics Still Stumps."
[10] Haldane, *Possible Worlds*, 286.
[11] Gombrich, *Sense of Order*, 9.
[12] Powers, *Orfeo*, 363.
[13] Strathern, *Mendeleyev's Dream*, 3-6.
[14] Taylor, "Newlands," 491-496, & Newlands, *Discovery*, 14-18.
[15] Strathern, *Mendeleyev's Dream*, 261.
[16] Powers, *Orfeo*, 31.
[17] Ohno, "All Pervasive," 71-78.
[18] Marden, *How They Succeeded*, 33.
[19] Einstein, *Autobiographical Notes*, in Calaprice, *Quotable Einstein*, 452.
[20] Langer, "Illusion of Control," 311.
[21] Hassin, *New Unconscious*, 82.
[22] Suzuki, *Zen Buddhism*, 97.
[23] Lippard, *Overlay*, 129-130.
[24] Isaacson, *Einstein*, 438.
[25] Isaacson, *Einstein*, 38.
[26] New York Times, "Sand Bars Trap Einstein."
[27] Lazar, "Meditation Experience," 1893–1897.
[28] Hölzel, "Mindfulness Practice," 36–43.
[29] Gilbertson, "Smaller Hippocampal Volume," 1242–1247.

30. Vermetten, "Long-Term Treatment," 693–702.
31. Koelsch, "Music-Evoked Emotions," 170–180.
32. Doyle, "Five Orange Pips," in *Adventures of Sherlock Holmes*, 225.
33. Konnikova, *Mastermind*, 26.
34. Thoreau, *Walden*, 144.
35. Thoreau, *Walden*, 146.
36. Cage, *Silence*, 93.
37. Herrnstein, *Bell Curve*, 307.
38. Kim, "Creativity Crisis," 285-295.
39. Schrobsdorff, "Teen Depression."
40. Huxley, *Perennial Philosophy*, 218.
41. Konnikova, *Mastermind*, 4.
42. Doyle, "Scandal in Bohemia," in *Adventures of Sherlock Holmes*, 159.
43. Sacks, *Musicophilia*, 48.
44. Huxley, *Perennial Philosophy*, 218.
45. Storr, *Music and the Mind*, 111.

F♯/G♭: Rausch

1. Csikszentmihalyi, *Flow*, 4.
2. Kotler, *Rise of Superman*, 21.
3. Kotler, *Art of Impossible*, 236.
4. Doyle, "The Second Stain," in *Adventures of Sherlock Holmes*, 684.
5. Fox, *Spirituality Named Compassion*, 129.
6. Cage, *Silence*, xxiv.
7. Carse, *Finite and Infinite Games*, 7-9.
8. James, *Energies of Men*, 15.
9. Csikszentmihalyi, *Flow*, 71.
10. Kounios, *Eureka Factor*, 160.
11. Reverberi, "Frontal Patients," 2882.
12. Sutherland, "Should they de-Nobel Moniz?"
13. Dietrich, "Neurocognitive Mechanisms," 746-761.
14. Lebovic, "Dionysian Politics," 2.
15. Lebovic, "Dionysian Politics," 2.
16. Schopenhauer, *World as Will and Representation*, 100, 261-262.
17. Page, "Unpretentious Prodigy."
18. Heraclitus, *Fragments*, 57.
19. Stravinsky, *Dialogues*, 39.

[20] Hill, *Rite of Spring*, 1.
[21] Stravinsky, *Conversations*, 19.
[22] Stravinsky, *Conversations*, 20.
[23] Gleick, *Genius*, 69.
[24] McCoy, "Otto Loewi," 3-4.
[25] Loewi, "Über Humorale," 629–640.
[26] Gray, *Henri Poincaré*, 216.
[27] Einstein to H. L. Gordon, (*AEA*, 1949), 58, 217.
[28] Brooker, *T.S. Eliot*, 138.

D♭: Atheoretical Cartography

[1] Cage, *Silence*, 79.
[2] Sontag, *Radical Will*, 11.
[3] Cage, *Silence*, 8.
[4] Cage, *Silence*, 12.
[5] Cage, *Silence*, 41.
[6] Cage, *Silence*, 26.
[7] Zuckerkandl, *Sound and Symbol*, 15.
[8] Storr, *Music and the Mind*, 168.
[9] Anderson, *Pianist's Craft*, xxvii.
[10] Rothstein, *Emblems of Mind*, 208.
[11] Parker, "History of Everyone, Ever."
[12] Orwell, *Nineteen Eighty-Four*, 35.
[13] Iliescu, "Impact of Artificial Intelligence," e24049.
[14] Hofstadter, *Gödel, Escher, Bach*, 677.
[15] Levitin, *Your Brain on Music*, 109.
[16] Levitin, *Your Brain on Music*, 93.
[17] Gleick, *Genius*, 323.
[18] Johnson, [Quoting Aristotle], "Number Theory in Music," 700.
[19] Brackenridge, "Kepler, Elliptical Orbits," 286.
[20] Brackenridge, "Kepler, Elliptical Orbits," 286.
[21] Horridge, "How Bees Distinguish Colors," 17-34.
[22] Bullock, "Radiant Heat Reception in Snakes," 19-29.
[23] Kalmijn, "Electric Sense of Sharks," 371–383.
[24] Lightman, *Accidental Universe*, 128.
[25] Rovelli, *Seven Brief Lessons*, 6.
[26] Rosado, "Gravitational Waves."

[27] Schlaug, "Corpus Callosum Size," 1047-1055.
[28] Burunat, "Action in Perception," e0138238.
[29] Vollman, "Plasticity Shapes Anatomical Properties," 245.
[30] Elbert, "Left Hand in String Players," 305-307.
[31] Doyle, "Sign of Four," in *Adventures of Sherlock Holmes*, 115.
[32] Lokhorst, "Fresh Light on an Old Codex," 293.
[33] Lokhorst, "Hemisphere Differences," 642.
[34] Van Wagenen, "Surgical Division," 740-759.
[35] Sperry, "Nobel Lecture."
[36] Sperry, "Disconnecting the Cerebral Hemispheres," 1223-1226.
[37] Sedbrook, "Molecules of Life."
[38] Browne, "Left-Handed Universe."
[39] Levitin, *Your Brain on Music*, 96.
[40] Storr, *Music and the Mind*, 169.
[41] Storr, *Music and the Mind*, 172.
[42] Taylor, *Stroke of Insight*, 143.
[43] Pesic, *Making of Modern Science*, 125.

Ab: Biomimicry

[1] Kobbé, "Wagner, Minna and Cosima," 263.
[2] Cage, *Silence*, 77.
[3] Richardson, *Marine Mammals and Noise*,
[4] Milius, "Music without Borders," 253.
[5] Doyle, "The Solitary Cyclist," in *Adventures of Sherlock Holmes*, 552.
[6] Cage, *Silence*, 113.
[7] Baptista, "Birdsong is Sometimes Like Music," 427.
[8] Shulman, "Program Notes."
[9] Powers, *Orfeo*, 113.
[10] Milius, "Music without Borders," 253.
[11] Baptista, "Birdsong is Sometimes Like Music," 426-443.
[12] Fehér, "Song Culture in the Zebra Finch," 564-568.
[13] Nottebohm, "Testosterone Triggers," 429-436.
[14] Young, *What the Robin Knows*, 55.
[15] Richards, *Peaceable Kingdom*, 73.
[16] Adeney, "European Starling."
[17] Baptista, "Birdsong is Sometimes Like Music," 436.
[18] Abert, *W. A. Mozart*, 727-728.

[19] Baptista, "Birdsong is Sometimes Like Music," 438.
[20] Planck, *Where is Science Going?* 217.

E♭: Semiotic Metaphor

[1] Dantzig, *Number*, 83.
[2] Sullivan, "Scientific Method," 275. (in Rothstein, *Emblems of Mind*, 151.)
[3] Rothstein, *Emblems of Mind*, 31.
[4] Storr, *Music and the Mind*, 182.
[5] Monk, *Ludwig Wittgenstein*, 44.
[6] Delacroix, *Journal*, 194.
[7] Dantzig, *Number*, 79.
[8] Phillpotts, *Shadow Passes*, 19.
[9] du Sautoy, "Listen by Numbers."
[10] Grinstein, *Encyclopedia of Mathematics Education*, 333.
[11] Rothstein, "Math and Music," sec. 2, pg. 1, 2nd to last par.
[12] Ehrenfest to Einstein, (*TCPAE*, 1921), 12, 30.
[13] Dantzig, *Number*, 44.
[14] Dantzig, *Number*, 106.
[15] Popper, *Conjectures and Refutations*, 123.
[16] Wasserstein, "Theaetetus," 165.
[17] Wasserstein, "Theaetetus," 165-179.
[18] Rothstein, *Emblems of Mind*, 34.
[19] Gingras, "Kepler's Harmonices Mundi," 259.
[20] Gingras, "Kepler's Harmonices Mundi," 259.
[21] Director, "Riemann for Anti-Dummies," part 15.
[22] Galileo, *Dialogues*, 95.
[23] Reston, *Galileo*, 7.
[24] Yanes, "Euler."
[25] Yanes, "Euler."
[26] Pesic, *Making of Modern Science*, 136.
[27] Yanes, "Euler."
[28] Dantzig, *Number*, 176.
[29] Rothstein, *Emblems of Mind*, 23.
[30] Fiore, "Music and Mathematics."
[31] Fiore, "Music and Mathematics."
[32] Mandelbrot, *Fractal Geometry of Nature*, 34.

33 Brothers, "Structural Scaling," 94.
34 Powers, *Orfeo*, 330.
35 Hofstadter, *Gödel, Escher, Bach*, 608.
36 Lightman, *Accidental Universe*, 7.
37 Einstein to Max Born, (*TCPAE*, 1926), 15, 426.
38 Poincaré, *Foundations of Science*, 44.
39 Rothstein, *Emblems of Mind*, 99.
40 Fiore, "Music and Mathematics."
41 Rothstein, "Math and Music."
42 Singh, "So-Called Fibonacci Numbers," 229-244.
43 Rizzi, S. "Fibonacci Sequence."
44 Roberts, *Music to Mathematics*, 183. (Cf. Lendvai, *Bartók*)
45 Delacroix, *Journal*, 194.
46 Wolff, *New Bach Reader*, 367.
47 Hofstadter, *Gödel, Escher, Bach*, 7.
48 Popper, *Conjectures and Refutations*, 95.
49 Hofstadter, *Gödel, Escher, Bach*, 79-81.
50 Zegers, "Eyes of Bach," 1429.
51 Hofstadter, *Gödel, Escher, Bach*, 86.
52 Hofstadter, *Le Ton Beau de Marot*, 65.
53 Hofstadter, *Le Ton Beau de Marot*, 65.
54 Isaacson, *Einstein*, 261-262.
55 Poincaré, *Foundations of Science*, 366.
56 Borroff, "George Crumb," 103-104.
57 Kant, *Critique of Judgment*, 247.
58 Hardy, *Mathematician's Apology*, 25.
59 Dyson, "Prof. Hermann Weyl," 458.
60 Lightman, *Accidental Universe*, 77.
61 Keats, "On a Grecian Urn," 149.
62 Clark, *Einstein*, 649.
63 Einstein overheard by Oswald Veblen, (*TCPAE*, 1921), 12, Introduction.
64 Planck to Einstein, (*TCPAE*, 1919), 9, 121.
65 Löwenstein, *Towards the Further Shore*, 156.
66 Einstein to Herbert S. Goldstein, (*AEA*, 1929), 33, 272.

B♭: Resonance

1. Einstein Trip Diary, (*AEA*, 1931) 29, 141.
2. Schopenhauer, *World as Will and Representation*, 312.
3. Mill, *Autobiography*, 144.
4. Mill, *Autobiography*, 145.
5. Isaacson, *Einstein*, 418.
6. Neville, *Won't You Be My Neighbor?*
7. Kafka, *Letters to Friends*, 16.
8. Avison, *Musical Expression*, 4.
9. Pais, *Subtle is the Lord*, 301.
10. Miller, "Evolution of Human Music," 356
11. Sperber, *Explaining Culture*, in Cross, "Music, Cognition," 35.
12. Levitin, *Your Brain on Music*, 259.
13. Levitin, *Your Brain on Music*, 191.
14. Nestler, "Cocaine Addiction," 4-10.
15. Casey, "William Styron."
16. Styron, *Darkness Visible*, 66.
17. Schopenhauer, *World as Will and Representation*, 264.
18. Reston, *Galileo*, 25.
19. Heilbron, *Dilemmas of an Upright Man*, 83.
20. Brown, *Planck*, 94.
21. Heilbron, *Dilemmas of an Upright Man*, 84.
22. Sacks, *Musicophilia*, 285.
23. Doyle, "Valley of Fear," in *Adventures of Sherlock Holmes*, 878.
24. Isaacson, *Einstein*, 161.
25. Doyle, "Sign of Four," in *Adventures of Sherlock Holmes*, 121.
26. Langer, *Philosophy in a New Key*, 221 – 222.

F: Synesthesia

1. Henahan, "Did Shostakovich Have a Secret?"
2. Sacks, *Musicophilia*, 69.
3. Jourdain, *Music, Brain, Ecstasy*, 172.
4. Maurer, *World of the Newborn*, 51.
5. Feynman, *What Do You Care*, 65.
6. Ward, "Synesthesia," 65.
7. Ward, "Synesthesia," 63.
8. Sacks, *Musicophilia*, 169.

9. Levitin, *Your Brain on Music*, 55.
10. Pesic, *Making of Modern Science*, 185.
11. Tu, "Seeing the Patterns of Sound."
12. Jung, *Memories, Dreams, Reflections*, 229.
13. McGuire, *Jung Speaking*, 275.
14. McGuire, *Jung Speaking*, 274.
15. Hyde, "Morphometry of the amusic brain," 2565.
16. Piccirilli, "Modularity of Music," 541.
17. Barnum, "Freud and Jung," 346.
18. McGuire, *Jung Speaking*, 274.
19. Nabokov, *Speak, Memory*, 35.
20. Nabokov, *Speak, Memory*, 35.
21. Sacks, *Musicophilia*, 252.
22. New England Journal of Medicine. "Cycling for Freezing Gait."
23. Sacks, *Musicophilia*, 254.
24. Pammer, "Brain Mechanisms and Reading Remediation," 1.
25. Forgeard, "Music and Phonological Processing," 383–390.
26. Overy, "Dyslexia, Temporal Processing and Music," 218-229.
27. Forgeard, "Music and Phonological Processing," 383.
28. Hoeft, "Long-Term Outcome in Dyslexia," 363.
29. Boatman, "Language Recovery after Left Hemispherectomy, 579-586.
30. Bogen, "Musical Tests," 524-525.
31. Marinkovic, "Right Hemisphere," 113.
32. Brust, "Music and Language," 367–392.
33. Taylor, *My Stroke of Insight*, 44.
34. Taylor, *My Stroke of Insight*, 174.
35. Raglio, "Improvement of Spontaneous Language," 240.

C: Apophany

1. Loy, *Musimathics*, 66.
2. Nagel, "Pythagorean Comma."
3. Halperin, "Quest for Bach's Temperament."
4. Clark, "Tuning Wars."
5. Clark, "Tuning Wars."
6. Clark, "Tuning Wars."
7. Lehman, "Bach's Extraordinary Temperament," 3-23.

8   Halperin, "Quest for Bach's Temperament."
9   Powers, *Orfeo*, 355.
10 Horgan, "Is the Schrödinger Equation True?"
11 Dantzig, Number, 176.
12 Rothstein, *Emblems of Mind*, 67.
13 Barrow, *World Within the World*, 348.
14 Popper, *Conjectures and Refutations*, 81.
15 Kuhn, *Essential Tension*, 342.
16 Horgan, "String Theory."
17 Nova. "A Radical Mind."
18 Rothstein, *Emblems of Mind*, 199.
19 Einstein, *Ideas and Opinions*, 274.
20 Einstein, "Geometry and Experience," (*TCPAE*, 1921), 7, 52.
21 Brooks, "There's a Glitch at the Edge of the Universe," Text Box.
22 Dantzig, *Number*, 236.
23 Weyl, *Philosophy of Mathematics*, 41.
24 Einstein, *Ideas and Opinions*, 28.
25 Horgan, "What Is Philosophy's Point?"

Finale

1   Storr, *Music and the Mind*, 174.
2   Kahneman, *Thinking, Fast and Slow*, 20-21.